W9-AVC-171

Venus & Mars

Influence Your Life and Relationships

ROBERT REID

Thorsons
An Imprint of HarperCollins*Publishers*

Thorsons
An Imprint of HarperCollins*Publishers*
77–85 Fulham Palace Road,
Hammersmith, London W6 8JB

First published in Australia by Mandarin 1996
Revised edition published by Thorsons 1998
3 5 7 9 10 8 6 4 2

A catalogue record for this book
is available from the British Library

ISBN 0 7225 3703 4

Printed and bound in Great Britain by
Caledonian International Book Manufacturing Ltd, Glasgow

Contents

Acknowledgements

While I have been working on this book over the last few years, many people have contributed their time, support and expertise. Thanks to Julie Austin, Helen Jenkins, Debbie Pepperdine, Sue Mackay, Ann-Maree Ellis, Ann-Marie James, Julian Lewis, Joan O'Shannassy, Trevor Gangmark, Julia Reid, Greg Sparkes, Deb Callaghan, Judith Davies, Judy Brookes, David Rosemeyer, Kirsty Low, Georgina Nebe, Kym Saddler and Alison Patten.

Also to the staff from Thorsons and HarperCollins, in particular James Holden, Natalia Link, Liz Hallam, Tim Nelson, Dominic Forbes, Hatty Madden and Michelle Pilley.

Introduction

Venus Signs and Mars Signs

You are probably familiar with your *Sun sign* – this is the sign occupied by the Sun when you were born. It is important in defining your personality in broad terms. What you may *not* know is that you also have a *Venus sign* and a *Mars sign*. These two signs are also very important in your life, but in a much more specific way than the Sun sign.

Your **Venus sign** indicates the way you share love and affection and the type of people you enjoy doing this with. It also describes your personal charm and beauty, and the way you express this in order to attract others.

For example, Marilyn Monroe and Shirley Temple were both born with Venus in Aries. Although different in many ways, both women are well remembered for expressing the innocent, child-like beauty associated with the placement of Venus in this sign.

Your **Mars sign** describes your passion and drive – what you want and the way you go about getting it – or *who* you want and the way you go about getting them. Mars is more self-centred and active than Venus. It is switched on, assertive and, as a rule, strongly sexual.

For example, although Rudolph Valentino and Richard Gere have different Sun signs, the on-screen sex appeal of these men is similar. Born when Mars was in the sensitive, emotional sign of Cancer, both men project a moody vulnerability which can be most attractive to women.

Venus is a *feminine* planet, associated with the goddess Aphrodite – into pleasure and erotic love. It is charming, sophisticated and sensitive to the needs of others. Mars is a *masculine* planet, associated with the god Ares – full of desire and passion. It is assertive and independent, challenging and playful. Each of us has these two sides to our nature. When fully expressed, they bring a healthy balance in relationships between togetherness and independence, creating a bond of love and friendship which still allows room for both individuals to be themselves.

Venus and Mars are both associated with sex, though in different ways. The sexuality of Venus is erotic and seductive, and conscious of the need for a mutually loving relationship. Mars, on the other hand, is more to do with passion, lust and conquest. As a rule, Mars obviously takes the lead and Venus responds, but this can vary according to the sign placement. For example, Venus tends to be more passionate and assertive in the fire signs – Aries, Leo and Sagittarius – while Mars is more subtle and seductive in the water signs – Cancer, Scorpio and Pisces.

In this book you will discover the meanings of your own Venus and Mars signs, as well as the Venus and Mars signs of others. You can also find out how well-matched your Venus sign is with the Mars signs of other people, as well as looking at the compatibility between your Mars sign and the Venus signs of others. These two compatibility readings offer you instant insight into the personal chemistry of your close relationships.

More on Venus and Mars Signs

You will notice that each chapter begins with a general description of the Venus or Mars sign, then continues with separate descriptions, one for a woman and the other for a man. This is because men and women tend to experience Venus and Mars in slightly different ways.

The average man will identify strongly with his Mars sign as a description of his 'male ego', i.e. the way he competes and asserts himself. Women, too, will relate to their Mars sign description, particularly when they are feeling strongly assertive, but they will also find their Mars sign describes qualities of strength in a man which they find particularly attractive.

... will identify closely with their Venus sign as

partner will identify more closely with their ... sensitive, socially aware partner will relate more strongly to their Venus sign.

While it is useful to make distinctions between Venus and Mars, and the roles commonly played by men and women in relationships, in reality it is not a black and white issue. Sometimes the traditional roles are completely reversed, in which case the woman will identify more strongly with her Mars sign while the man feels more attuned to his Venus sign. In fact, most people should relate to both descriptions to a greater or lesser extent. Make sure you read both.

There are other factors in the individual birthchart which have a bearing on passion, pleasure and the personal side of relationships generally. For example, aspects between certain planets, and Sun and Moon signs. But Venus and Mars signs are particularly important, and a good place to begin. I hope this book will bring you pleasure and enjoyment, as well as some insight into the way you relate to others.

Compatibility

In astrology, the Venus-Mars connection is seen as one of the most powerful indicators of personal attraction between the sexes, so obviously you will want to know how your Venus and Mars signs connect with those of other people. Compatibility readings are located after the descriptions of the Venus and Mars sign in each chapter. For each relationship there are *two* compatibility readings – one between your Venus sign and the other person's Mars sign, and another between your Mars sign and their Venus sign.

In describing the nature of different Venus-Mars connections, these readings will help you understand why you click so easily with certain people – particularly partners or potential partners. The relationship between Venus and Mars is very intimate, and when it is strong there will be a high level of personal chemistry between yourself and the other person.

If you are the sort of person who takes an interest in delving more deeply into your inner life, you will find it worthwhile to look at the compatibility readings between your *own* Mars and Venus signs. This will give you an insight into your different Mars-Venus needs in personal relationships. You may find these needs are quite similar. On the other hand, you might discover two distinctly different sides to your nature which you seek to fulfil through close contact with other people.

Finding Your Way Around this Book

ephemeris[1]) to locate the Venus and Mars positions for your birth year.

The columns on the left side of the page show the sign positions of Mars as they change from month to month; the columns on the right side show the changing sign positions of Venus.

Locate the sign positions of Venus and Mars corresponding to your date of birth, and make a note of them.

2 Look up the Mars and Venus signs of your partner, lover, close friend, or current fantasy, and make a note of these also.
3 Turn to the chapter on your Mars sign for a description of your Mars personality, then refer to the compatibility readings to check the compatibility between your Mars sign and other people's Venus signs.
4 Turn to the chapter on your Venus sign to read about your personality in love and friendship, and your compatibility with other people's Mars signs.
5 If you were born on the last day that Mars (or Venus) spends in a sign, you will find that this day is also listed as the first day that Mars (or Venus) spends in the following sign. In this case you are said to be born on the 'cusp' of the two signs. For example, if you were born on April 14 1956, you will find this given as the last day Mars spent in Capricorn *and* the first day it entered Aquarius. In this case you should read the chapter for Mars in Capricorn as well as the chapter for Mars in Aquarius. People born with planets on

the cusp tend to be a mixture of both signs. However, depending on your time and place of birth, you may find you relate more strongly to one description than the other. If you are a cusp person and you are left feeling the need for further clarification, then perhaps it's time you took the extra step of having your chart drawn up and fully interpreted by a professional astrologer.

[1] The ephemeris is based on Greenwich Mean Time (Great Britain). For places west of Greenwich (e.g. Canada, the USA) the cusp will sometimes include the second last day the planet spends in a sign. For places east of Greenwich (e.g. South Africa, Australia) the cusp will sometimes include the second day after the planet enters a new sign.

PART 1

What's Your Passion?

1

Mars in Aries

Who but me, when but now?

B.B. TYRWHITT-DRAKE

With Mars in Aries there is a very wilful and self-centred side to your nature. Once your passions are aroused you feel an eagerness to act, and you are not afraid to take the initiative. Inspired by the excitement of the moment, you can be very spontaneous. Some may call you impetuous, rash, even inconsiderate, but you'll generally be too busy getting on with things to notice. Anyway, even if you *are* a bit thoughtless at times, you are not intentionally hurtful. When challenged, you can usually claim innocence, and in the meantime you've managed to get things done.

Having Mars in Aries gives you a powerful independent streak. You may assert this in a physical way, or you could be the independent intellectual. In either case you have a strong sense of your individuality and will express this in whatever field you choose to enter – like Jack Nicklaus *(The Bear)* and Greg Norman *(The Shark)*. Or Queen Victoria, who left the stamp of her identity on an entire century.

You are not inclined to seek the approval of others before acting out your desires or expressing your opinion. You want to be free to explore new avenues of thought and activity, doing your own thing your own way in your own time. *I Did it My Way* could be your theme song. You find unnecessary restriction and limitation to be galling. In fact, even *necessary* restriction and limitation is something you would prefer to avoid.

When interacting with others, you enjoy the challenge of competition – for you, it adds spice and excitement. Vanessa Redgrave described this quality in the actor Franco Nero with whom she had a long relationship, 'He played every game – soccer, pinball and table

[illegible]

cially competitive sport, is the best way to work off excess energy.

Aries is naturally a childlike sign, and, like the other two fire signs (Leo and Sagittarius), very playful. With Mars in Aries, mucking around, acting-up like a big kid, and generally being rude, crude and boisterous will stir your passions and turn you on.

In ancient Egypt, the Ram was associated with sexual potency. Having Mars in Aries, you will be particularly turned on by sexual conquests as a confirmation of this power, but you may lose interest after the initial excitement dies. In a steady relationship, you will look to find new ways of making sex stimulating – as if it's the first time every time. You probably wouldn't want to go as far as the Marquis de Sade, who had a very exciting sex life but spent half his time in jail because of it – but a certain element of risk and adventure will always add spice. This is part of the attraction of Paul Newman, who has been said to act with 'a high promise of sex and danger'.

The Woman with Mars in Aries

This is a symbol of the warrior woman. The main Greek warrior goddess was called Athene. She was born from Zeus's head when it was split open with an axe. However, although she had the physical strength and courage of a warrior, she more often used her head when dealing with conflict. So while you may express this side of your nature in a physical way – say, through physical exercise and sport – you may also be a strong-minded intellectual. Even if you choose to maintain a softer image, behind that feminine charm there is a hard edge to your character which will come out whenever you are challenged. When you feel comfortable displaying your assertive power, you will be an openly independent, highly competitive woman, prepared to fight for what you want and not afraid to go it alone.

As a woman with Mars in Aries you may be tempted to dominate the man in your life, but if he allows you to do this he won't really turn you on. In fact, you will be most passionately attracted to a man who is independent and assertive with plenty of initiative – a man who knows his own mind and does his own thing. If he has a few rough edges you won't really mind – they go with the character.

He may even be quite self-centred, but this doesn't matter – you are turned on by the strong, fearless type, and someone who is too nice and considerate just won't stir your passions. Even Queen Victoria, supposedly the ultimate in fuddy duddies, was susceptible throughout her life to 'strong men with roguish charm'.

The Man with Mars in Aries

Mars and Aries are both associated with masculine power and potency, so having Mars in Aries can make you very blunt and assertive. Even if other factors in your chart modify your expression of this, when you really want something, or someone, you will still adopt a direct, no-nonsense approach. Sometimes you may be too abrupt and undiplomatic but, within reason, many will find your open, honest energy to be disarmingly attractive.

As a Mars in Aries man, you also have a strong independent streak. When you are really switched on you will express your own opinions, make up your own mind and go your own way, and if others don't like it that's their bad luck. You're not passionate about popularity, just getting what you want.

You may focus your energy into one of several different areas. Perhaps you are the independent intellectual, pursuing your own ideas and theories. On the other hand, you may be very physically active, and into competitive sport. Whatever your domain, you have a very

[illegible]

With Venus in Aries

When you meet someone with Venus in Aries the personal attraction between you will be very strong. This is a feisty contact – very passionate and playful. You will raise each other's energy levels – like two kids egging each other on in a competitive but pleasurable way. Even when you fight it can be enjoyable, as both of you thrive on a certain amount of conflict. It will be natural for you to take the lead, and you should find this person will respond with pleasure. If you are close, sex will be very energetic and lustful.

With Venus in Taurus

Although the Venus in Taurus person prefers a slower, more comfortable pace, they will still appreciate the value of your strong, assertive energy if they can see it is going to lead to something productive. Just remember to direct some of this energy into making sure they are comfortable. The Venus in Taurus person enjoys being pampered. If they like you, they will enrich your relationship in a pleasurable, sensual way. In return, you will bring excitement into their life. On a personal level, this contact is very passionate and physical.

With Venus in Gemini
This is an easily compatible pair. The Venus in Gemini person will value your high energy level, especially if it is directed into activities which demonstrate your skill and cleverness. In this case, they will encourage you to explore new directions and develop your talents even further. If you look like becoming overheated, their detached response will help you keep a cool head. In return, you could bring some fire and passion into their life. This person enjoys talking, so make sure you know where their intellectual interests lie.

With Venus in Cancer
This is a challenging contact. The attraction can be very strong, but there is likely to be some friction between you. The Venus in Cancer person values a soft, sensitive approach so you had better stop and think before you rush in too hard. Be kind and considerate. They won't respond until they feel secure, no matter how sincere you are. If this person likes you, they will be very caring and attentive to your needs. In return, you could bring lust and passion into their life. When it works, this is a very stimulating connection between two quite different people.

With Venus in Leo
This is a strong, easy connection. You may find that Venus in Leo tries to tie you down (Leo is big on loyalty), but it will be done in such a pleasurable way that you probably won't mind. In fact, with a bit of firm direction you may find you start to really achieve something with this person. The contact is warm, playful and high-spirited. If you are close, sex will be powerfully passionate. Just make sure you tell this person how wonderful they are on a regular basis.

With Venus in Virgo
This is a bit of an odd couple. Although you are both independent in your own way, the Venus in Virgo person is much more introverted and analytical, and is more likely to criticize than encourage you. However, if you can handle this you may find it useful, and it will lead

you to direct your energy with greater care and precision. In return, you can bring excitement and adventure into this person's life. If you are looking for intimacy, pay special attention to cleanliness.

ing up on your social [illegible]
you will be able to coax this gentle, sophisticated person into your lair.

With Venus in Scorpio

Both Aries and Scorpio are passionate signs, but Scorpio is secretive by nature so this person is unlikely to respond quickly to your frank, direct approach. Have you ever considered subtlety? If the Venus in Scorpio person likes you, and you can tune in to their moods, they will give you strong and loyal support. In return, you could inspire them to take a few risks and be more outgoing. When it works, this contact has a high emotional charge. Sexually, it could be very passionate.

With Venus in Sagittarius

Aries and Sagittarius are two of the most adventurous signs, so the connection here is strong and easy. Prepare for a roller coaster ride. Perhaps you should take out some accident insurance before you really get going! This can be a very passionate contact. If you are on the same wavelength you should be able to really let go with this person. Whether you play sport together, travel together or just lust after each other constantly, you will feed each others' energy and have lots of fun.

With Venus in Capricorn

Have you ever considered going into business? Perhaps a sports shop or a travel agency? The Venus in Capricorn person will encourage you to direct your powerful, passionate energy into something constructive. The connection here can be difficult, but it will be very stimulating. This person won't love you unless you can make something of your life, but on the other hand you could inspire them to let go and take more risks. If neither of you is prepared to give, there will be conflict, but if you like each other enough you should be able to resolve this, and your differences will be a source of stimulation. Perhaps bedroom therapy would help.

With Venus in Aquarius

Aquarius is a freedom-loving sign like Aries, so the connection here is an easy one. The Venus in Aquarius person will respond to your open, honest approach with friendly affection, especially if you have something to say which is worth listening to. This person values intellectual friendship, and they will encourage you to keep your cool and communicate clearly. In return, you might inspire them to open up in a warm and passionate way.

With Venus in Pisces

Pisces and Aries are very different signs. Usually this person is looking for someone with a much softer approach than yours, but if you can direct your high-powered energies in a sensitive way, they might just respond with affection. This person has a very sympathetic and caring side to their nature. If they really like you they will be sensitive to your feelings and attentive to your needs. In return you could bring some excitement and adventure into their life. On an intimate level, this contact is full of feeling and passion.

Famous Personalities with Mars in Aries

Oliver Cromwell
Marquis de Sade
Alan Arkin
Jack Nicklaus
Paul Simon

Greg Norman
David Letterman
Sarah Miles
Pete Townsend
Kurt Russell

2
Mars in Taurus

Firmness is that admirable quality in ourselves
that is detestable stubbornness in others.
ANON

Yeah, right. Check out the list of celebrities at the end of the chapter for some seriously stubborn people.

With Mars in Taurus you have great persistence and determination, and a down-to-earth approach which will help you achieve the results you desire. You don't make a big noise about it – Taurus is a practical sign and has no need of a fanfare – you just put your head down and get on with it. Your style is slow but sure. Once you make up your mind you really want something, you settle down for the long haul, working hard and never giving up until you reach your chosen goal. If you are thwarted in the short-term, you have the patience to wait, for years if necessary, until the time arrives when you can realize your ambition and get what you really want. Like the pilot Douglas Bader, who lost both legs in a plane crash but never gave up hope of flying again. It took him seven years to get his licence back. He then went on to become one of the top fighter pilots of World War II.

After you have achieved your objective you are capable of hanging on with great tenacity – no one is going to take away what you have gained through long, hard work. This doesn't mean you can't be generous – just so long as it's your choice and on your own terms.

You will be attracted to material wealth, initially because you are a practical person who wants stability, security and comfort. As you are naturally a hard worker, you are unlikely to stop there and, given time, ... substantial wealth – like 'Material

...

if they let you. And if your relationship ...
become extremely jealous – jealousy and possessiveness go hand in hand. As a rule, you will find that people don't like being seen as your property. Funny about that. How flexible you are with these matters will depend on other factors in your chart.

Your sex drive is strong, but controlled. You need to be sure you really want someone before you make a move. Even then you don't get turned on in a rush – the pace is definitely slow but, like a bull, once moving you are very difficult to deflect from your course. When your passions are aroused, you are extremely sensual, with a powerful desire for physical contact.

The Woman with Mars in Taurus

Although Mars is essentially a masculine energy, Taurus is one of the feminine signs, so you will probably feel quite comfortable with this placement. Your passion is very earthy and sensual, and you are well tuned in to the practical necessities of survival. The quiet nature of your determination may fool some people into thinking you're a pushover, but if they take you on they'll find out they're wrong. Bulls are *not* to be trifled with.

You will be strongly attracted to a man with strength and determination – someone who knows where he stands, knows what he wants and won't give in until he gets it. This man has staying power. He may try to dominate you, and you might even let him – for a while at least.

You desire a man who is earthy and sensual. Sex is important to you on a purely physical level. You'll also be attracted to a man who has plenty of money, partly because of the opportunity to own things of value and beauty, but also because you desire stability, security and comfort. Alternatively, you may be attracted to the artistic type, but he would need to create a wealth of beauty to hold you on a permanent basis. Whatever form they take, power and wealth turn you on.

The Man with Mars in Taurus

As a man with Mars in Taurus you are quietly determined, if not downright stubborn. When you assert yourself you do so with great strength, knowing exactly where you stand and refusing to be pushed around. Like the bull, you are slow to anger, but if someone tries to take what's yours they'll find out how powerful your temper can be.

Before you make your move, you need to feel you are in a strong position. No one is going to rush you into anything you don't want, or feel unprepared for. As Taurus is a fixed sign, you are capable of building up enormous power, whether this is material power or simply the power of your own determination. So if it comes to a showdown, and you are ready, you'll usually win. You are strongly independent and self-contained but, if you choose to team up with someone, your solid strength and reliability are excellent qualities in a relationship. Staying power is something you possess in great measure. In marriage, you like to see yourself as a good provider.

Your Compatibility with the Twelve Venus Signs

How will that special person respond to your Taurean approach? Before reading on, turn to the back of the book and look up his or her Venus sign. Then read the relevant section below.

With Venus in Aries

In relationships this person is very independent, and they enjoy the company of someone with a playful, adventurous spirit. You may find they are not as stable as you want them to be, but their high-spirited

will value your earthy practicality, and your constancy and reliability will help them feel pleasantly comfortable. They will tune in to your desire to create wealth, and their appreciation and support will encourage you to be even more productive. If you are close, sharing physical pleasures with this person will be a joy. Together you could indulge yourselves in the rich sensuality of life, whether in the art gallery, the garden, the kitchen or the bedroom. This is a very earthy contact, and the physical pleasure of sex will be very strong.

With Venus in Gemini

The Venus in Gemini person enjoys a faster pace, so you may have to speed up a bit here. I know you're not inclined to prance about, but how about walking quickly? They also enjoy socializing, and like to have a variety of friends. If you want to win their affection, it helps to be flexible and communicative. If this person likes you, they will liven up your life with new experiences and interesting conversation. In return, you could bring comfort and stability into their life.

With Venus in Cancer

The contact here is easy, and moderately strong. This is definitely a stay-at-home combination, though it could also work well in business. The Venus in Cancer person has strong motherly tendencies, and you should be quite happy to be pampered. Perhaps they will give your neck a massage while you put your feet up after dinner. But first you might have to cook the meal – or at least pay for it. The possibilities for mutual indulgence are endless.

With Venus in Leo

This is a strong contact, but Leo and Taurus are very different signs so be prepared for something different. The one thing Leo and Taurus have in common is they're both fixed. This person may be marginally less stubborn than you are, but not enough to give in all the time. You both need to show some flexibility for this one to work. While *you* want to stay in your comfortable, secure rut, the Venus in Leo person will be more inclined to party and have fun. They probably won't be averse to throwing your money around either – shock, horror! On a more positive note, this person is extremely loyal, and if you can resolve your differences you may have a very stimulating and passionate relationship.

With Venus in Virgo

This is a strong and easy contact. The Venus in Virgo person will value your down-to-earth approach. They may be a bit fussy and critical, but if they like you they will express this with affection and you will probably find it useful. This can be an excellent close, working relationship. It may also be a strong physical and sexual one. Perhaps the best news for you is that Venus in Virgo is not only practical but flexible – this means they are willing to fit in with someone else – isn't that a bonus?

With Venus in Libra

While your approach is very basic and earthy, this person relates to others in a detached, intellectual way. They enjoy the company of people who are gentle, courteous and fair-minded, so if you want them to like you, you'd better polish up on your manners and social etiquette. This person could help you develop a softer, more diplomatic approach in all your relationships. In return you might inspire them to express themselves with greater strength and determination. If you are looking for intimacy, start with interesting conversation.

With Venus in Scorpio

This is a very powerful connection, but there is likely to be some tension and conflict. The attraction is most definitely there, but whether [illegible] will depend on your ability to see things

With Venus in Sagittarius

This person is attracted to someone who is philosophical by nature and extravagant in taste – doesn't really sound like you, does it? Still, if you feel like something completely different, this could be just the right person. They will encourage you to break from your familiar routine and explore new experiences. If you are open to it, this person could bring a new sense of excitement into your life along with fun and laughter. As a compromise, perhaps you could travel to exotic new places in comfort and luxury.

With Venus in Capricorn

This is a strong, easy contact. The Venus in Capricorn person will value your down-to-earth approach, encouraging you to show initiative. Whether your relationship is business or personal, they will work hard to help you achieve your goals, adding their own touch of class along the way. This connection could develop slowly but surely into an extremely productive partnership. On an intimate level, the contact is very earthy and physical.

With Venus in Aquarius

This contact can be strong, but Taurus and Aquarius are very different signs so there is bound to be a certain amount of conflict. This person likes to socialize, and values people who are friendly and communicative. They are also very free-spirited, and won't appreciate it if you try to own them or tie them down. If you want them to like you, try cultivating a more detached approach and discuss what you want rather than seeking to impose your will. When it works this can be a very stimulating contact between two quite different people, but it requires effort and compromise from both.

With Venus in Pisces

The connection here is fairly easy and moderately strong. While you are fixed, this person is flexible and, if they like you, willing to fit in – isn't that good news? You can offer them a sense of stability and firm direction in their life, and in return they will be very attentive to your needs. They may even persuade you to be more sympathetic towards other people's opinions. Essentially this is a physical and emotional bond. When things are going well it will be extremely comfortable.

Famous Personalities with Mars in Taurus

George Sand
Tchaikovsky
W.C. Fields
Charlie Chaplin
Adolf Hitler
Salvador Dali
Bette Davis
David Niven
Douglas Bader
Ginger Rogers
Lucille Ball
John F. Kennedy
Liberace
Sidney Poitier
Shirley MacLaine
Rudolf Nureyev
Muhammad Ali
Mick Jagger
Madonna
Billy Joel
Michael Jackson
Eva Peron
Tom Cruise
Robert de Niro
Paula Abdul
Belinda Carlisle
Olivia Hussey
Belinda Jackson
Carole King
Margot Fonteyne
Peter Fonda
Roy Orbison
Jamie Lee Curtis
Prince Edward
Christine Keeler
Gwyneth Paltrow

3
Mars in Gemini

If other people are going to talk, conversation is simply ...

ANON

With Mars in Gemini, you have a passion for communication. Your strong, restless, nervous energy needs an outlet, and the more fired up you are, the more you can talk – and aren't there so many things to talk about! As Dustin Hoffman said of Meryl Streep 'She eats words for breakfast.' Queen Elizabeth I was another person with a passion for words. She expressed herself with great intelligence, and also had a razor-edged wit.

For you, variety is definitely the spice of life, and you have the ability to focus on several subjects at once, moving rapidly backwards and forwards between your chosen topics. When your mind is clear and your focus sure, you are able to weave a fascinating intellectual mosaic – fast, aggressive and entertaining, full of interesting information and clever, witty arguments which will dazzle your audience. Your mind is quick and versatile, and you are good at adapting to the ever-changing circumstances of your immediate environment.

Negatively, your desire for new stimulation may cause you to keep skipping from one idea to the next or from one project to the next. It's important for you to have a focus which really captures your interest, otherwise you may expend a great deal of energy but the results will be only superficial.

Being the passionate intellectual, you enjoy debating and arguing. If you work at this you will develop a high level of skill. Sometimes, though, you may be too quick to turn a conversation into an argument.

As well as talking, you may have a passion for writing. Your strong curiosity and desire to learn may also make you an avid reader. Games, jokes, tricks and puzzles will captivate your mind, offering further variety and stimulation.

On a more physical level, Gemini rules the hands, and these offer yet another outlet for the expression of your nervous energy. Apart from gesturing while you talk, you may develop a high level of manual skill in a particular area. Perhaps you play a musical instrument, or maybe you are into sport. The more speed and versatility required, the more the activity will appeal to you.

Sometimes, if you are totally bored and there is nothing to do, you might just hop in the car and go for a drive, or jump on your bike and go for ride. Shopping, visiting, whatever – just moving around and checking things out is interesting.

When it comes to sex you need variety. A comfortable routine will soon bore you and then you will seek new experiences to maintain your interest. Positively, this means you can be very inventive, forever finding new approaches and discovering different games to play. For example, Hugh Grant's on-screen character is said to reflect his own personality, projecting a strong sense of sexuality which is 'unpredictable and various'. Curiously enough, Gemini rules transport in general, cars in particular (I'm not sure about BMWs).

With Mars in Gemini, words and sex go together naturally, so you should find any form of erotic communication to be particularly stimulating. Elizabeth I may have been the 'virgin queen', but she still loved the foreplay of passionate letters and ribald jokes. James Joyce was another person who took great pleasure from lusty letter writing. Much of what he wrote is unprintable here. Perhaps, if he were alive today, he would have taken equal pleasure from phone sex.

The Woman with Mars in Gemini

You are a sociable and articulate person with a strong desire to com-
... and you will never be short of

...

so you will always seek change and variety to keep ...
For an ideal balance you will have interests requiring physical skill, manual dexterity and, most importantly, an alert and active mind.

You are attracted to a man you see as being skilful and intelligent. He may be an excellent craftsman, or perhaps he's more the academic type. Either way, he has a strong desire to learn and the ability to express himself in a way which you find exciting. You will need to keep your mind in good shape to match his wit or hold your end up in a lively debate. Above all, this man is interesting – he may even have a touch of genius in him. One thing is certain – you won't be bored.

The Man with Mars in Gemini

You assert yourself by seeking to impress on others the level of your knowledge and skill. When you feel confident you express yourself intelligently and articulately, and what you say is worth listening to. You are capable of speaking with great passion and you make a very strong debater. Despite this passion you can maintain a clear, detached awareness, guiding your mind through the maze of information and ideas which are relevant to your subject. Sometimes you are overly argumentative; at other times your mind may hop around too much, making it difficult to follow your path of reasoning. However, when you are able to maintain a clear, detached focus and the right sense of proportion you will be a most impressive speaker and others will stop to hear what you have to say.

Your Compatibility with the Twelve Venus Signs

How will that special person respond to your Gemini approach? Before reading on, turn to the back of the book and look up his or her Venus sign. Then read the relevant section below.

With Venus in Aries

This is a high-energy contact between two different but compatible signs. In relationships, this person is not as rational and detached as you are, but like you they enjoy living for the moment and you should find their feisty, energetic manner pleasantly stimulating. You may play sport together, or perhaps you prefer intellectual games. Either way, this person enjoys competition, and may also have a mischievous streak. If your games progress to the bedroom, you will find they can also be very passionate and lustful.

With Venus in Taurus

The Venus in Taurus person enjoys the company of someone who is reliable and down-to-earth. In relationships they are very sensual, and have a strong appreciation of life's physical pleasures. If you want this person to like you, you will need to tone down your nervous, restless energy and pay attention to their need for comfort. If you have an artistic bent they should value this and encourage you to be productive. In return you can brighten up the relationship with your wit and humour.

With Venus in Gemini

This could be a very stimulating friendship. The Venus in Gemini person should enjoy your quick wit and lively banter, responding with equal speed and agility. Their detached, friendly replies will soften your energy and encourage you to express yourself with more grace and eloquence. They love variety, and when it comes to sharing pleasure and good times you should be the most inventive couple on the block. At social gatherings the two of you can help to make things more lively and interesting, inspiring others to open up and have fun. If you are

... variety, and when you are

...

the mood they will quietly withdraw. ...

be sensitive to your feelings and attentive to your needs. In return, you could distract them out of their bad moods with some light, friendly banter.

With Venus in Leo

This can be a very upbeat contact between two people who enjoy having fun. In relationships, the Venus in Leo person is warm-hearted and playful. They will like you most of all when you are confident and outgoing; then their enthusiastic response will inspire you to even greater heights of creative activity, and you will keep them amused and entertained in a most pleasurable way. This contact can be full of games and laughter, and very theatrical. Just remember to treat this person with respect, and don't hold back on the praise and compliments they so enjoy.

With Venus in Virgo

In some ways these signs are quite similar, but in other ways altogether different. The Venus in Virgo person enjoys the company of someone who is thoughtful and practical. They may find the things you say interesting, but their response will often be critical, especially if they don't think your ideas are leading somewhere useful. Don't bombard this person with aimless chatter. Before you speak, think seriously about what you want to say so that you can present it clearly. This person will challenge you to be more thoughtful and self-critical. If you handle this well, they will warm to you, and then you will find they are most attentive to your needs.

With Venus in Libra

This is a strong, easy contact. The Venus in Libra person will appreciate your detached, intellectual approach, responding in a polite and thoughtful way. They will encourage you to be balanced in your thoughts and to always treat people fairly. You should pick up a few tips on diplomacy from this person, and they may also introduce you to some interesting new people. Together you make a very sociable pair, never short of something to talk about and good at keeping the conversation flowing at dinner parties or other social events. If you are close, the conversation should continue pleasurably into the bedroom.

With Venus in Scorpio

This is a bit of an odd couple. If you have the sort of mind which looks beneath the surface of everyday life, the Venus in Scorpio person will be most attentive. They have no interest in trivialities, so don't bother with superficial chatter. If you want to win their affection, be prepared to talk about your feelings, but don't expect an instant response – this person is deeply sensitive and will only open up to someone they really trust. However, when they *do* open up, they are extremely passionate, and also very loyal and committed.

With Venus in Sagittarius

Gemini and Sagittarius are opposite signs, but opposites attract, and when this connection works it is powerful and stimulating – just be prepared for a challenge. The Venus in Sagittarius person will be attracted by your open, communicative approach. However, they like someone who has a vision and looks to the future, so if you become too caught up in logical details and can't see beyond the present moment they may become restless and bored. When the contact works well, you will find this person inspiring and fun to be with. Together you could travel, explore new ideas and play games. When it comes to sex, this connection has a strong emotional charge.

With Venus in Capricorn

This person likes someone who is strong, down-to-earth and sure of where they are going. If you want to win their affection you will need to do more than entertain them with interesting ideas. What is your plan, [illegible]

This is a strong, harmonious contact. [illegible] will appreciate your detached, intellectual approach and will respond with their own ideas. They are more fixed than you, so they will encourage you to focus on a particular topic and talk it through. Together you could share many friendly, stimulating conversations. You will also enjoy an active social life – in fact, this person could introduce you to a variety of interesting people. If you are close, your conversation will be more intimate, but it is unlikely that you will ever stop talking.

With Venus in Pisces

This contact can be very stimulating, but as Pisces and Gemini are different in nature there is likely to be some conflict. The Venus in Pisces person enjoys the company of someone who is soft and sensitive to feelings. If you are too detached and intellectual they won't like it. You need to make an effort to get in touch with their mood before you start talking. Slow down, take it easy, put on some pleasant music. If this person really likes you they will help you relax, give your brain a rest, and float off to somewhere pleasurable. In return you could help them clarify their thoughts when they are feeling confused.

Famous Personalities with Mars in Gemini

Queen Elizabeth I
Beethoven
Charlotte Brontë
Thomas Hardy
F. Scott Fitzgerald
Louis Armstrong
Jimmy Stewart
Dean Martin
Sean Connery
Julie Christie
Barbra Streisand
Linda Evans
Meryl Streep
Arnold Schwarzenegger
Andre Agassi
Gabriela Sabatini
Wesley Snipes
Hugh Grant
Eva Braun
Coco Chanel
Catherine Deneuve
Phyllis Diller
Princess Margaret
Billie Jean King
Camilla Parker Bowles
Diana Ross
Uma Thurman
Camille Claudel
Benito Mussolini
Jim Morrison
Julio Iglesias
Erica Jong
Carlos Santana
Peter O'Toole
Lindsay Wagner
Antonio Banderas
Virginia Woolf
Prince Phillip
Kate Winslet
Tony Blair

4
Mars in Cancer

There is something about holding on to things that [illegible]

EDNA O'BRIEN

You are a deeply emotional person with sensitive feelings and powerful moods. When you are feeling good you will be very sensitive and caring, but if you feel depressed you can be extremely crabby and irritable. When your passions are aroused you may be overcome with emotion, whether this is tears, joy or anger. In a difficult situation you may try to withdraw rather than risk giving vent to such powerful feelings. Sometimes this is a wise move, but as a rule it is important for you to make your feelings clear.

Mars in Cancer can be quite an emotional razor-edge. Famous personalities with highly strung feelings include Al Capone, Pablo Picasso and Rudolph Valentino who could 'turn the air blue with profanity when his temper was aroused'. You may also have a tendency to sulk when things don't go your way.

You are fiercely attached to people with whom you feel a strong, emotional bond, and very protective of those you care about. These feelings may extend generally to the family, but are felt most strongly for those you trust and are close to. If there is any conflict within the family you will be very sensitive to this and it will give rise to passionate feelings within you. Again, you will protect those you care for.

It is often not recognized, but Cancer is a very enterprising sign. You have the ability to lay the foundation of a business, working closely with a group of trusted people. This could be a commercial business, a family, or just a group of people who share a dream.

Your strong feelings of attachment may extend beyond close friends and family to your local community, state or country – whichever one you feel defines your place in the world. You have a deep desire to experience a sense of belonging, to know your roots. This may extend to a passion for history, whether this is your family history or, more broadly, the history of your culture or religion.

Check out the list at the end of the chapter for people who have had a strong attachment to home, family, country and/or history. Whether they have sung about it (Tom Jones), written about it (Shakespeare), acted it out (Kenneth Branagh, Emma Thompson, James Cagney), or revisited it (Lord Byron, Isadora Duncan), their passion is emotionally driven.

As a Mars in Cancer person you can be very soulful, with feelings which run so deep that their origins are a mystery. You may be drawn strongly to certain people or places without knowing why. You may also experience a psychic attunement with those people you feel close to. Isadora Duncan described such connections with the men in her life.

Your sexuality is very strong and powerfully emotional – there's nothing detached about Mars in Cancer. You need to feel safe and secure with the other person before you let go, but when you do, your feelings are very intense. Your combination of strength and emotional sensitivity can be extremely seductive. Such is the case with Rudolph Valentino and Richard Gere, whose on-screen sex appeal lies in their moody vulnerability.

The Woman with Mars in Cancer

You express your moods and feelings with great power and passion, and there is a strong motherly side to your nature. Obviously, if you have children, these feelings will be directed towards them – you will feel very attached to them emotionally and psychically, and will be extremely caring and protective. But these feelings will also be extended towards anyone you really care about, including family pets.

You are strongly attracted to a man who is sensitive, caring and pro- [illegible] But you also want him to need

[illegible]

As a man you have a deep, emotional sensitivity, and can assert yourself with great feeling, but you need to be comfortable with this type of expression to use it effectively. While Mars symbolizes the traditionally masculine approach – i.e. independent and assertive – Cancer is the sign most concerned with emotional safety and security. There is a dilemma here. If you assert yourself too strongly, you risk alienating the people you are close to – and you don't want to do that. But if you are too concerned with keeping the emotional peace, you won't be assertive enough, and after a while your frustration will probably boil over into anger. To get the right balance here you need to be sure about what you want and where you stand with others so that you can express yourself clearly and sensitively.

You have strong initiative, and the ability to establish the foundation of a business or family. You are very caring and supportive towards those you trust, but if they betray you they will be confronted by your hard, crusty shell.

Your Compatibility with the Twelve Venus Signs

How will that special person respond to your Cancerian approach? Before reading on, turn to the back of the book and look up his or her Venus sign. Then read the relevant section below.

With Venus in Aries

This is a powerful connection, but the energies are quite different so the contact is challenging. While you become strongly attached and emotionally involved, the Venus in Aries person is quite playful and independent in relationships. This doesn't mean they're insincere, just more detached, with a mischievous sense of fun which you might not always appreciate. However, other factors in your chart may incline you to be more open to their free-spirited energy, in which case, when the mood is right, you will find their company pleasurably stimulating. If you are close, your connection will be extremely passionate.

With Venus in Taurus

This is a cosy contact, and quite a strong one. The Venus in Taurus person is very sensual in relationships. They should share your love of security, and if you get on well you could settle into a warm and pleasurable routine of mutual self-indulgence. While you create a cosy atmosphere full of warmth and feeling, they will attend to the physical pleasures and comforts. A well-stocked pantry, some comfortable armchairs and a luxurious bed should keep you happy for weeks.

With Venus in Gemini

The Venus in Gemini person enjoys moving around, socializing and talking a lot. As long as they don't move around too much to threaten your security, that will be okay by you. In fact, they could lighten your moods and provide interesting entertainment and stimulating conversation. In return, you can offer them a sense of emotional security, and introduce them to the world of powerful feelings you know so well. If you are looking for intimacy, make sure you know where their intellectual interests lie.

With Venus in Cancer

This is a powerful combination. If you feel secure together, you will connect on a deep, emotional level with a contact which is very sensitive, and at times psychic. The Venus in Cancer person will pick up on your moods and respond in sympathy. When you are feeling down they will nurture and care for you, when you feel threatened they will protect you, and when you feel good they will share your pleasure, adding their own loving touch. Is this too good to be true? When it

close, you will find this person can be very passionate and lustful.

With Venus in Virgo

This is a complementary pair. The Venus in Virgo person will value your quiet strength and commitment. Their common sense practicality provides a natural foil for your moodiness, and if they like you they will be quite happy to work hard at helping you achieve your goals. You take the initiative and this person should fit in. If they criticize, it is only because they want to make things better. On an intimate level this could work well. Just make sure the sheets are ironed.

With Venus in Libra

The connection here can be quite stimulating, but the energies are very different and it will be a challenge to make the relationship work. The Venus in Libra person values detached, intellectual contact. They will probably ask you to explain your feelings and you'll at least need to make a reasonable effort. If you succeed, they will be fascinated; if not, they will probably think you're a bit crazy. While you are deep, mysterious and primitive, this person likes relationships which are refined and civilized. Perhaps the best way to express your feelings to them would be through a song or a poem.

With Venus in Scorpio

This is a strong, emotional connection. The Venus in Scorpio person will value your sensitivity and your passionate expression of feelings, and they should respond to your moods with sympathetic understanding and affection. Like you, this person is powerfully emotional, so making love would generate a very strong bond of feeling. If you care deeply for each other, don't be surprised to find your connection is psychic. This could be a mutually seductive relationship.

With Venus in Sagittarius

The Venus in Sagittarius person loves to discover new ideas and places, so don't expect an enthusiastic response to your offer of emotional attachment and security. If you want this person to like you, try to cultivate your sense of humour and adventure. If you could afford to have several homes dotted around the world it would help. Maybe you could start a travel agency which specialized in historical tours? On an intimate level, the contact is passionate and emotional.

With Venus in Capricorn

The attraction between these signs is strong, but there is also a certain amount of tension – opposites either complement or fight. Like you, the Venus in Capricorn person is looking for security in a relationship, but they value the practical, hardworking approach and won't necessarily sympathize with your moods. However, they should appreciate your initiative and commitment – like you they want to build something, whether this is a personal relationship or a business. When things are going well they will help you keep your feet on the ground while you bring a sense of mystery and romance into their life. If you are close, there will be a strong sexual connection.

With Venus in Aquarius

The Venus in Aquarius person values someone who is detached, friendly and intellectual – doesn't sound much like you, does it? It would be worth your while to develop a more cool-headed approach with this person. Try to think clearly and communicate honestly, even when you're in a rotten mood. Also, don't expect them to hang around all the time – this person values their freedom and enjoys having a variety of friends. If they really like you they will encourage you to [illegible] you could help them

[illegible]

if you get uptight about something or someone [illegible]
love with this person is more an emotional than a physical experience. If you are close, they should respond readily to your seductive nature. Together you could share many pleasurable fantasies.

Famous Personalities with Mars in Cancer

- Nostradamus
- William Shakespeare
- Marie Antoinette
- Mozart
- Lord Byron
- Isadora Duncan
- D.H. Lawrence
- Rudolph Valentino
- Al Capone
- James Cagney
- Burt Lancaster
- Robert Mitcham
- Tony Curtis
- Audrey Hepburn
- Natalie Wood
- Tom Jones
- Liza Minnelli
- Hayley Mills
- Cyndi Lauper
- Diane Keaton
- Mikhail Gorbachev
- Richard Gere
- Nastassja Kinski
- Kenneth Branagh
- Emma Thompson
- Isabelle Adjani
- Ingrid Bergman
- Sheena Easton
- Anjelica Houston
- Diana Rigg
- Viviene Leigh
- Leonard Nimoy
- Pablo Picasso
- Keanu Reeves

5
Mars in Leo

Everyone has a right to my opinion.

ANON

You are turned on by a sense of your own personal power, and get a real buzz out of expressing this with confidence and style. When things are going your way, you are warm, outgoing and fun to be with – very upfront, with strong, positive energy.

Once you decide you really want something, or someone, you can apply yourself with tremendous commitment and enthusiasm, and others will find it hard to resist your warm, infectious energy. You are so positive you just carry people along.

So what's the down side to this wonderful person? Well, quite frankly, you may suffer from vanity. 'Who's suffering?' you say, 'I'm justifiably proud of my talents.' Right, okay. You can also be very bossy and domineering – but quite possibly you see this as another one of your strengths. During the filming of one of her movies, Jodie Foster was nicknamed BLT – Bossy Little Thing – and she didn't mind a bit.

As Leo is a sign which celebrates its own power, you are naturally self-centred. When your passions are aroused it seems as if the whole world revolves around you – or if it doesn't, then it should. You have the qualities of a born actor and are turned on by being in the lime-light. Like a child, you are a bit of a show-off, and enjoy being the cen-tre of attention. You want to be loved, admired and respected. If you are not given the centre of attention when you want it, you can be childish and demanding. You might even indulge yourself in a display of melodrama. If someone is downright disrespectful and really pushes your buttons, you can become very angry – and won't they know it.

Your sex drive is likely to be strong. For you,

playful and above all a celebration of your personal power. Naturally, you will want to take the lead. If your partner is inclined to shower you with compliments, and praise your beauty and performance, you won't mind a bit. In fact, open admiration turns you on and helps you generate even more warmth and general good vibes. A good example is the singer Edith Piaf. Although she was small in stature – her name is French slang for 'sparrow' – her sex drive was powerful. She was very generous to all her lovers, buying them wardrobes of clothes, and offering financial assistance when they needed it.

The Woman with Mars in Leo

Mars in Leo is a strongly masculine energy. As a woman you may feel comfortable expressing this, but you may also choose to tone it down a bit. Even if you aren't loud and extroverted, others will still feel the strength of your character and know you are not to be trifled with. You want to have things your own way and will follow your desire with great determination. You can be a lot of fun to be with – very warm and playful. You are a bit bossy too, and not afraid to put your foot down when you really want something. Leo is a fixed sign, so it is stubborn.

For a man to turn you on, he needs to have a big strong personality to match yours. Any sign of weakness or pettiness will turn you right off. You want someone who is grand and confident, the sort of man who inspires respect from others, and is warm, noble and generous – and if he's *that* good, you won't even mind taking a backseat for a while. Edith Piaf had many relationships, but her one true love was the graceful and muscular Arab-French prize-fighter, Marcel Cerdan.

The Man with Mars in Leo

You get a kick out of impressing others with the power of your personality. If you could, you would be king, and then you could tell everyone what to do. But you'd want to be a popular king, because then you would get plenty of praise and admiration. So as well as being noble and magnificent, you are also warm and generous to your loyal subjects, bestowing expensive gifts and acts of kindness on those most worthy. Of course, the disloyal and unworthy get short shrift, and perhaps a dose of the royal temper. And if someone makes the mistake of causing you humiliation, you may choose to banish them from the kingdom on a permanent basis – or at least until they've grovelled sufficiently.

Your Compatibility with the Twelve Venus Signs

How will that special person respond to your Leonine approach? Before reading on, turn to the back of the book and look up his or her Venus sign. Then read the relevant section below.

With Venus in Aries

This is a strong, easy contact – very fiery and upbeat. The Venus in Aries person should appreciate your warm, enthusiastic approach, and if they like you their response will be immediate and positive. Sometimes this person may be a bit too independent for your liking, but if you basically get on well there is a playful chemistry here which will keep drawing you back together. Maybe you play sport together, or perhaps your games are more theatrical. On an intimate level, this contact is extremely passionate and lustful.

[illegible]

sex. In return, you could inspire them to be more [illegible] bringing passion and lust into their life.

With Venus in Gemini

The Venus in Gemini person should appreciate your high, positive energy levels, and could provide you with a variety of entertaining responses. However, you will have to stay on your toes, because this person likes to keep things moving and will quickly become bored if you get stuck in one of your domineering ruts. Still, the connection here is quite complementary, so this person should be able to divert you in a friendly and interesting way. In return, you will feed them with warm, passionate energy. This can be quite a creative combination.

With Venus in Cancer

The Venus in Cancer person likes sharing a relationship with someone who is soft and caring, and sensitive to their moods and feelings. Are you listening to this? If you are strong and insistent you may beat this person into submission, but they won't enjoy it. While you are confidently moving on to your next favourite topic they will be quietly slipping away. But if they really *do* care for you, they may just disappear to cook you a meal or run you a bath. If you want to woo this person, you will need to take a more sensitive approach. When things are going well you will bring them out of their shell and make them laugh, while they help you get more in touch with the world of inner personal feelings.

With Venus in Leo

This is a very powerful, passionate contact. The Venus in Leo person is attracted to your confident, self-centred energy. They should respond quickly and with pleasure, and when you happen to be the centre of attention they will gladly share this, adding their own grace and style. If you want to play king, this person is the perfect queen. They love the style, the glamour and the glitter, and when you are at your warm, noble best, they will love you too. A couple of royal peas in a royal pod.

With Venus in Virgo

When this person responds to your full-on, dominating style, it could well be with a sharp, critical comment. Even though your ego may not like it, you would do well to listen. The fact that they have bothered to respond at all shows they like you, and if you follow their advice you may find your performance improves. The Venus in Virgo person values someone who pays attention to detail and is prepared to work at getting something just right. In return they will respond in a dutiful and practical way. This person could teach you a thing or two if you are prepared to stop and listen.

With Venus in Libra

This person values someone who is cool, charming and sophisticated. While you may qualify as charming, there's a definite question mark over your level of cool and sophistication. Still, there is enough in common here for you to bluff your way through. Just keep the humour flowing with style, and remember to make a really big effort to listen to someone else's opinion every now and then. If you can express a social conscience, or show an active interest in art, this will be a bonus. Essentially, the connection is complementary, and you should be able to enjoy each other's company without too much difficulty.

With Venus in Scorpio

There is a strong attraction here, but the energies are very different and there could well be friction. While you are open and straightforward in your approach, the Venus in Scorpio person is more reserved when it comes to sharing love and affection, so don't expect them to respond with obvious enthusiasm. First they need to trust you, and this will

[illegible]

the gods have blessed your relationship. The Venus in Sagittarius person will value your honest, energetic style and should respond with open enthusiasm. This person has a sense of fun and a spirit of adventure, and enjoys sharing these in a relationship. When you are feeling good they will encourage you to show it, and if you are feeling down they can jolly you out of yourself. This could be a pleasurable, passionate and uncomplicated relationship.

With Venus in Capricorn

The Venus in Capricorn person will value your strength and power – as long as it's not all show. If you are bluffing, and they find out, they will be decidedly unimpressed. To earn their love and friendship you need to show you are a person of substance. They value hard work and commitment in relationships and take great pleasure in sharing this load with someone they care for. If you measure up to their judgement, this person will help you create a very productive relationship, and in return you could inspire them with your warm, positive energy.

With Venus in Aquarius

This contact can be very stimulating, but as Leo and Aquarius are opposite signs there is likely to be some conflict. The Venus in Aquarius person will value your clear, direct approach, and should respond with equal honesty. Unfortunately, they won't take your ego into account, so you mightn't always want to hear what they've got to say.

But while this person may be cool and detached, they are also genuinely friendly. In fact, they can help you keep your head when you are overheating and in danger of blowing a fuse. If you really like them, you will find their cool, friendly response is the perfect complement to your passionate approach. On an intimate level this contact has a high emotional charge.

With Venus in Pisces

Pisces and Leo are very different signs, so the contact here is not an obvious one. The Venus in Pisces person values someone who is soft and sensitive. They also like people who can rise above their own personal interest and direct their energies towards some higher cause – helping others for instance. If you want this person to enjoy your company you will need to make a big effort to be more sensitive to feelings and less self-centred. In return, you will find this person to be a very caring and devoted partner. When it comes to sex, they can be very seductive, and also quite imaginative.

Famous Personalities with Mars in Leo

Hans Christian Andersen
Dostoyevsky
Mata Hari
Gary Cooper
Gregory Peck
Sophia Loren
Brigitte Bardot
Paul McCartney
Goldie Hawn
Cher
Michael J. Fox
Florence Nightingale
Harrison Ford
Jodie Foster
Hillary Clinton
Michael Jordon
Edith Piaf
Robert Redford
Ringo Starr
Candice Bergen
Gina Lollobrigida
Imelda Marcos
Bette Midler
Sylvia Plath
Claudia Schiffer
Sigourney Weaver
Victoria Abril
Joanna Lumley
James Dean
Boy George

James Taylor
Neil Young
Richard Dreyfuss
Ray Davies
Pierre Trudeau
Demi Moore

6

Mars in Virgo

Oh, wouldn't the world seem dull and flat
with nothing whatever to grumble at.

W.S. GILBERT

You approach things in a careful, considerate way, studying every detail and looking at a situation from different angles. In fact, you have quite a passion for analysis – sifting, sorting and turning things over in your mind, rearranging the pieces and touching up the details. Whatever it is that consumes you, whether at work or in a relationship, you will keep to the task until everything is in its proper place. I think the word is 'perfectionist'.

In your quest for perfection, you have no doubt learnt at an early age that it helps to be systematic. If you find an approach that works, whether it's an efficient way of getting someone into bed, or a new, more streamlined method of assembling washing machines, you will patent it for re-use. Why waste precious time starting all over again when you've already got it worked out?

But then again, like all good perfectionists, you realize deep down that the ultimate task you have set yourself is impossible to achieve. There is always some small detail that's not quite right, and of course, the smaller it is the more it annoys you. So you are bound to want to fiddle and make changes. Doing this will work up to a point, but if you don't know when to leave things alone you may become uptight and worry too much. If you get angry, you will express this in a cold, critical way.

Whenever you get enthusiastic about something, or someone, your strong critical faculty comes to the fore, like the actor Sophie Marceau, who is 'usually ready with a spiky opinion'. You may appear overly

active interest in health. After all, if you don't have a healthy body and a healthy mind your performance will suffer, so anything you can do to improve things here will help. You may develop a passion for diet or fitness, or perhaps you take an interest in improving the efficiency of your mind. Whichever course takes your fancy, you will find there are many different avenues to follow. The challenge is to develop a focus and pursue it efficiently and effectively.

King Henry VIII of England was heavily into physical fitness, and in his prime was said to be able to tire out four or more horses in one day. More recent examples of fitness fanatics include Sylvester Stallone and the actor Michael Landon who had a full gymnasium installed at every house he owned. Princess Diana was also an exercise fanatic. I don't know whether the writer Mary Shelley was herself into physical health and fitness, but her novel *Frankenstein* tells the story of a medical student who attempted to create the perfect human being, free of all disease. Obviously it was an issue for her.

When it comes to sex you are more likely to be matter-of-fact than romantic. Mars in Virgo has a reputation for sexual modesty which may extend to prudishness. Such was the case with the writer Ernest Hemingway. On the other hand, Clark Gable was quite promiscuous. This man definitely had charisma. As Joan Blondell put it, 'He affected all females, unless they were dead.' But he avoided complicated romantic entanglements, saying, 'I do not want to be the world's greatest lover.' This down-to-earth sexuality of Virgo is less confronting than the high romanticism of some other signs which often promise more than they can deliver.

The Woman with Mars in Virgo

You have a fussy and discriminating side to your nature, and a strong, analytical mind which you can apply in your work and relationships. You work hard at improving your abilities, but you may be inclined to worry too much when things are not working out. When this happens you need to be able to relax and let go so that you can return to the task with renewed focus and clarity. Although you may be highly critical in relationships, you also have a strong desire to help people with their problems.

You will be attracted to a man with a strong, practical, analytical mind. While you may criticize each other, if done in a healthy spirit your relationship is bound to improve and together you could achieve a great deal. Big-talking idealists won't turn you on – you are much more impressed by someone with a discerning intellect, a strong sense of duty and a fine eye for detail. This man is intelligent and thoughtful, self-contained and hard-working. He has great inner strength, and strives for perfection in all that he does.

The Man with Mars in Virgo

You assert yourself most effectively when your mind is clear and you have a well-grounded understanding of the situation in all its detail. Using your powers of critical analysis, you take a problem-solving approach and present your solution with clarity and precision. No matter what the other person throws at you, you have the answer, and you can back it up with powerful, complex arguments. After all, you have already taken the time to analyse your own position carefully and critically. Your trump card is simply this – your way works best, and if challenged you will be able to explain why. You have no need to beat others into submission or resort to clever tricks and manipulation – and besides, if it works well then everyone stands to benefit, don't they?

With Venus in Aries

The Venus in Aries person values someone who is bright and spontaneous, so you will need to lighten up a bit if you want to win this person's affection. By all means be assertive, even critical, but if you try to get them involved in a long, serious analytical discussion, don't expect them to respond with pleasure. If this person really likes you they will encourage you to take a more open, playful approach. In return, you could inspire them to be more thoughtful and discerning.

With Venus in Taurus

This is a strong, easy contact. The Venus in Taurus person will value your down-to-earth approach, and their practical response should help you focus your energy and become even more productive. While you initiate activities in an efficient, well-organized way, they could work with you to create something which is pleasurable and beautiful, as well as useful. This person has a strong appreciation of life's physical pleasures. If you are close, your connection will be very sensual and physical.

With Venus in Gemini

This contact is very strong, but Gemini and Virgo are quite different signs so there is bound to be some friction. Like you, the Venus in Gemini person has a very active mind, but they prefer light, witty conversation to heavy analytical discussion. You may find their chatty, friendly replies to be more distracting than helpful, but then again, perhaps you're getting too serious. Choose your subjects carefully, and if you know they are not interested, don't bother. When you *do* click, the contact will be very stimulating and you will have plenty to talk about.

With Venus in Cancer

The quiet sensitivity of this person can nicely complement your serious, practical style. Both Cancer and Virgo are very caring signs, so you should be attentive to each other's needs. If they like you, the Venus in Cancer person will be sensitive to your feelings – very supportive and protective. In return, you could help them organize their life in very practical ways. This is a good home-making combination. If you are close, your relationship will be strongly physical and emotional, and very private.

With Venus in Leo

The Venus in Leo person enjoys the company of someone who is confident and outgoing. While you may be confident, they will enjoy your company all the more if you express yourself with a bit of flair and style. If they really like you, this person will encourage you to be less critical and more playful. You might need to bite your tongue and overlook the odd annoying detail. In fact, if you find this person attractive you would be wise to tell them so. Buy them flowers, make them feel special. On an intimate level this connection is very passionate and physical.

With Venus in Virgo

This is a very strong, harmonious combination. The Venus in Virgo person will appreciate your practical, analytical approach, and even if they respond critically it will probably be done with affection. When you are uptight about something, they will try to help you find a solution. You will both have a strong sense of duty to the relationship and to each other, working together closely and efficiently. Whatever you create, this person will add their own touch of beauty, paying particular attention to the details. When it comes to sex, the connection is strong – very physical and sensual.

With Venus in Libra

The Venus in Libra person enjoys socializing, meeting new people and exchanging ideas. They like someone with a clear, balanced mind

With Venus in Scorpio

The Venus in Scorpio person should appreciate your quiet, practical, thorough approach. Virgo and Scorpio are both serious, private signs, so the contact can be very close and intimate. If you *are* close to this person it will be a deeply caring relationship, and very self-sufficient, with little need of the outside world. Your common sense approach will naturally balance their powerful moodiness. This person has a strong sexuality and can be extremely seductive. On an intimate level your connection would be very sensual and full of feeling.

With Venus in Sagittarius

The person with Venus in Sagittarius likes someone who is openly passionate about life and enthusiastic about discovering the world. They may have a philosophical bent, or perhaps they are into sport or travel. Obviously Virgo and Sagittarius are very different signs, so you will need to make an effort to bridge the gap. While you are working hard on the details, they prefer discovering the big picture. If you know broadly what interests them, perhaps you could amuse them by providing a detailed analysis – they may even find it useful. Just remember to deliver it with humour and feeling. In return, this person might help you find a new and exciting direction in your life – perhaps one filled with lust and passion.

With Venus in Capricorn

This contact is strong and harmonious. The Venus in Capricorn person will value your down-to-earth approach, and should enjoy working with you to create a strong relationship. If you have a problem they will be only too happy to help you resolve it, and you will find their no-nonsense attitude to be useful if not inspiring. Whenever you take the initiative this person's response will encourage you to build something of lasting value and beauty. Once the work has been done, you should find it easy to share the rewards of pleasure. If you are close, your contact will be very physical and sensual.

With Venus in Aquarius

Like you, the Venus in Aquarius person has a strong, serious mind. However, while you present yourself as self-contained and inward-looking, this person is much more socially outgoing and enjoys the company of a variety of different people. If they really like you, they will encourage you to be more sociable – perhaps to join a club or a group where your practical skills could be of service. This can be a stimulating, intellectual contact between two quite different people.

With Venus in Pisces

The connection here can be very strong and stimulating. Although Virgo and Pisces are opposite in nature, they may complement each other perfectly. This person has a softness and emotional sensitivity which can balance your practical, critical approach. They will encourage you to be more easy-going and sympathetic, while you challenge them to be more down-to-earth and discriminating. You will have your differences, but in a way these make the attraction stronger. Sexually, this could be quite a potent contact.

Famous Personalities with Mars in Virgo

Henry VIII

Jacqueline Kennedy Onassis
Kim Novak
Yoko Ono
Brooke Shields
Michael Caine
Princess Diana
Sylvester Stallone
Stevie Nicks
Michael Landon
Rosanna Arquette
Olivia De Havilland
Elizabeth Hurley

Indira Gandhi

Alice Cooper
Evel Knievel
Charles Bronson
Barbara Cartland
Jerry Garcia
Bob Geldof
Charlton Heston
Dorothy Parker
Fidel Castro
Robin Williams
Helen Hunt
Sophie Marceau

7
Mars in Libra

I'm giving you a definite maybe.
SAM GOLDWYN

With Mars in Libra you have the most aggressive, self-centred planet in a sign which is extremely polite and considerate of others. You may express this in two different ways:

When Libra prevails, you are very courteous and co-operative in your relationships. You want to work closely in a partnership and are prepared to align your energy with the other person's – together you can accomplish more than you would as individuals. This is your modest side – very thoughtful and well-mannered. Although you desire to have your own way, you see this in terms of what is best for the relationship, so you are able to put a lot of energy into making sure the relationship works.

On the other hand, when Mars prevails, relationships bring out your sense of aggression and competition. In this case, arguing and debating with others excites you. You find the conflict stimulating, and when it is not there you become bored. When your partner puts forward a point of view you feel a passionate desire to take the other side. It's not that you don't like the other person but, after all, Libra is the sign of balance between two opposites, so you are just doing your bit to maintain the balance. If you feel particularly bored, you may even invite conflict by saying something you know will stir the other person to argue with you. This makes the relationship more exciting, and if the other person takes it well and gets into the spirit of debate your contact will be enjoyable and thought-provoking. As long as your disputes are [illegible] respectful they will clear the air, stir your passion and

[illegible]

your relationships [illegible]
quite detached. Even when you are debating passionately, [illegible] still respond to a well-reasoned argument from the other person. In fact, well-reasoned arguments turn you on. If you argue too much you might alienate people, but if you are too soft and polite you may allow others to take advantage of you. When you strike the right balance between Mars and Libra, you will be strong and assertive, but also polite and reasonable.

Your desire for equality and balance in relationships may give rise to a strong social conscience. In this case you will relish taking part in public debate over issues which fire your passion. John Lennon certainly did this. A more modern example is Roger Moore who is an enthusiastic special ambassador for UNICEF. If you have a partner to share this interest with, all the better – but you won't necessarily agree with everything your partner says, and your private debates will provide further stimulation.

Sexually, you are fairly detached, and may prefer to direct your passions into 'higher' intellectual pursuits – George Bernard Shaw said, 'Intellect is as much a passion as sex.' When you do get down to it you will probably be quite formal and polite – a very thoughtful, civilized sexuality, where respectful communication is all-important.

The Woman with Mars in Libra

You have a strong desire for balanced, harmonious relationships, and you are always prepared to sit down and talk things through to achieve this end. If someone offends your natural sense of fairness you will want to put matters right by explaining clearly and logically why you disagree with their point of view. Even when it is not in your interest, you may still uphold a position simply because you believe it is just.

You are attracted to a man who is intelligent, articulate and fair-minded. He may have a strong social conscience and a desire to participate in public affairs. In your personal relationship he will direct a lot of energy into discussing things of interest to you both. This man is a gentleman in the true sense of the word. He is sensitive and refined, with a passion for fairness and justice.

The Man with Mars in Libra

You assert yourself in a detached, rational way, winning over others through the power of reason. You need to be sure of your own mind before you can do this effectively, so you will want to think through all the arguments before you speak out. When you *do* have your say, you may express yourself in a polite and diplomatic way, carefully explaining your reasons to the other person. On the other hand, you might argue your point aggressively and enjoy the ensuing debate. The most important thing is to have a clear understanding of your position and why you hold it. In this case, even when you feel very passionate about something you will maintain a clear, detached perspective. You may also argue well on behalf of others.

Mars in Libra brings to mind the classic, old-fashioned image of the gentleman – perhaps on the conservative side, but able to treat everyone in a fair-minded and respectful manner. Roger Moore has this side to his nature, as did Elvis Presley.

Your Compatibility with the Twelve Venus Signs

How will that special person respond to your Libran approach? Before

his or her Venus

your sense of

Libra type. In contrast to your intellectual detachment, this person

a fiery lover – very impulsive, with a mischievous sense of fun. As long as they keep within socially acceptable boundaries you may find them very attractive. They will draw you out of yourself and encourage you to be more passionate and spontaneous. This can be quite a potent sexual contact.

With Venus in Taurus

Taurus and Libra may share an appreciation of artistic beauty, but they are quite different signs. While *you* like to maintain a certain level of detachment in relationships, the Venus in Taurus person tends to become strongly attached and may also be possessive. They will be attracted to your refined sensibility, but don't expect them to respond to well-reasoned arguments. As a rule, if they don't like something they won't budge. However, if you get on, they will help you become more focused and productive as well as sharing their rich appreciation of the physical pleasures of life. This person makes a very sensual lover.

With Venus in Gemini

This is a strong, easy contact. Like you, the Venus in Gemini person enjoys relating on an intellectual level, so when you initiate discussion they could well respond with pleasure. Whatever subject you get passionate about, this person is likely to join in – their active mind can pick up on a theme and take it in a variety of different directions. No doubt you will find this very stimulating. If you are close, you will have many interesting and enjoyable conversations which will continue on a more intimate level in the bedroom.

With Venus in Cancer

This person is unlikely to respond warmly to your detached, intellectual approach, so you will need to come down out of the clouds if you are going to make an impression here. However, if you *do* connect it can be very stimulating. The Venus in Cancer person likes someone who is sensitive to their moods and feelings. Once settled in a relationship, they enjoy the homely routines of life, and won't necessarily be as interested in pursuing new social contacts as you are. Perhaps you could compromise by entertaining interesting people at home.

With Venus in Leo

The Venus in Leo person will enjoy your company when you are strong, confident and outgoing. If you are sure of what you want to say, then say it boldly and with style. And if you can inject some humour, then all the better. Like you, this person enjoys the company of others. They may be a bit self-centred, but they make up for this with their natural warmth and exuberance. If you find them attractive, let them know they are special. When it comes to intimacy, this person can be very passionate and playful.

With Venus in Virgo

The Venus in Virgo person will value your active mind, but will encourage you to apply it in a more practical way. When you get carried away discussing ideas, they will probably respond critically to bring you down to earth. If there is an earthy side to your own chart, you will value this, if not, it may irritate you. When you *are* being practical, this person will enjoy working hard to serve your interests, paying particular attention to detail. They have a strong sense of duty and can be very caring in a relationship.

With Venus in Libra

You could establish a very close relationship with this person, based on good conversation. They will value your intellectual approach and,

With Venus in Scorpio

Don't expect this person to respond readily to your open, detached approach. In relationships, the Venus in Scorpio person is deeply emotional and very private. They will only be drawn into intimacy if they trust you – constancy and loyalty mean more to them than clever ideas and refined conversation. You need to talk to this person about your feelings. If they really like you, you will find out what deep, passionate relating is all about. This person has a strong sexual nature, and knows how to enjoy the more primitive pleasures of life.

With Venus in Sagittarius

This person should value your strong, intellectual energy, responding with warmth and enthusiasm. Like you, they enjoy exercising the mind, but their thoughts are more abstract and inspirational, while you tend to be very rational. When you present your carefully reasoned point of view, their response may be quite passionate. This is a philosophical person with a sense of humour who enjoys sharing new experiences. The contact is easy and pleasurable, and it may give you a whole new perspective on yourself in relationships.

With Venus in Capricorn

This person values someone who knows where they stand, takes charge and gets on with things. When you are vacillating and generally finding it hard to make a decision, keep your mouth shut, otherwise they will probably start yawning. If you show weakness or indecisiveness, you are more likely to receive a harsh judgement than a sympathetic response. When it works, this contact can be very stimulating, but the signs are quite different so a special effort is required. When you are strong and clear-headed, and your ideas are practical, this person will listen to what you have to say and then work hard to help you in your endeavours. They may be practical and judgemental, but if you measure up you will find they are very supportive.

With Venus in Aquarius

The connection here is strong and easy. This person will value your detached, rational approach, and if you are finding it difficult to make up your mind they could help you come to a decision. Like you, they enjoy social interaction, and their active mind ensures they always have something to talk about. They are open to progressive ideas, and take pleasure in debating these in an honest and reasonable way. This could be a very stimulating relationship based on friendship and shared intellectual interests. If you are close, your conversation will lead naturally to intimacy.

With Venus in Pisces

You may share a sense of idealism with this person, but your detached, rational approach won't always strike a chord – it depends on their mood. When your words create the right atmosphere, they will respond with affection, but you need to be relaxed and in touch with your feelings. If this person really likes you, they will help you to calm down when your mind is buzzing with unresolved thoughts – possibly through gentle touch or perhaps by putting on the right music. Even though you have your differences, this can be a very gentle, caring contact. On an intimate level, this person can be very seductive and imaginative.

Famous Personalities with Mars in Libra

Jonathan Swift
Johnny Carson
Peter Falk
Elvis Presley
John Lennon
David Carradine
Michael Douglas
Cybill Shepherd
Roger Moore
Cliff Richard
Roseanne
Bill Clinton

Andy Warhol
Anais Nin
Herb Alpert
Dalai Lama
Edgar Allen Poe
Peter Gabriel
Whitney Houston
Nicole Kidman
Emmanuelle Beart
John Malkovich

8
Mars in Scorpio

What does not destroy me makes me stronger.

NIETZSCHE

Mars in Scorpio is a symbol of quiet, persistent power. You don't make a big deal of it, but others will pick up on your energy – which can be quite magnetic – and recognize your underlying strength. Robert Kennedy, Grace Kelly and Jimi Hendrix are three examples of famous people who possessed such a powerful emotional aura.

Although you won't necessarily show it on the surface, you are a person with very strong feelings who becomes deeply attached to whatever, or whomever, you desire. Developing this attachment may take time, but once you know what you want, you set your course with quiet determination and pursue your chosen goal with great thoroughness and purpose. Obstacles will not deter you, in fact, if anything, they will make you more determined than ever. In order to get what you want, you are prepared to endure difficulties which would defeat a more faint-hearted soul.

The essence of your power is emotional, and you are naturally secretive. When someone blocks your course, you will try to skilfully manoeuvre the situation to get your own way. This secrecy may be the hallmark of a clever and subtle tactician who works modestly behind the scenes. On the other hand, you may simply be ruthless and manipulative. Scorpio is a sign of extremes – of saints and sinners. You may be driven by fear and jealousy, trying to control everyone who comes into your life, or you might be a wonderful person who is motivated by a deep love for your fellow human beings. It is quite possible [illegible] both of these characters within yourself. How you deal

[illegible]

and sarcastic [illegible] offends you deeply it will take a long time for them to regain your trust – if at all.

Because of your capacity for deep attachment, you may sometimes be afflicted by jealousy. The French author Colette wrote about this emotion in a very evocative way, saying that it bloomed inside her 'like a dark carnation'. Deep red is one of Scorpio's colours.

Your sex drive is strong, though you are capable of self-denial if it suits your purpose. You are likely to be secretive about this side of your life too – who knows what you get up to in the privacy of your bedroom, or what secret fantasies occupy your mind? One thing is certain, sex needs to be intense so that your deepest feelings are aroused.

Mahatma Gandhi was preoccupied with sex during his early years. At the age of 37 he became celibate, and said that his passion was reinforced and raised to a higher spiritual level. Scorpio is one of the most primitive signs, but is also associated with the process of inner transformation.

The Woman with Mars in Scorpio

You have deep, strong feelings which others will pick up on and respond to. Whether you are comfortable with this will depend on other factors in your chart as well as the way you were brought up. Your energy is powerfully emotional and sexual, and you have a great deal of personal magnetism. As soon as you want something, or someone, you will start transmitting your desire. You are probably quite aware of this, but you're not letting on, right? Mars in Scorpio prefers to work behind the scenes, unnoticed. You can tap in to the lines of power, and if you're smart, use them to your advantage.

You will be attracted to a man with quiet, inner emotional strength – someone who knows what he wants in life and follows his path with thoroughness and determination. He may be a bit on the obsessive side, but this is an indication of his strength of purpose – once committed he is not easily put off. You trust him to remain loyal – that's important to you – and together you share a deep bond of intimacy which is very private and full of emotional intensity.

The Man with Mars in Scorpio

Some people may interpret your quiet confidence as an indication that you are an easy-going sort of person. Well, you may be easygoing when nothing is at stake, but when your passion is aroused and you know what you want, you pursue your goal with great purpose and determination. In fact, when the chips are down you are capable of being quite ruthless with yourself as well as others, willing to endure difficulty and hardship to achieve your goal. The key to your power is the emotional intensity with which you focus your energy. You are also fiercely protective of the people and things which are important to you.

Your Compatibility with the Twelve Venus Signs

How will that special person respond to your Scorpio approach? Before reading on, turn to the back of the book and look up his or her Venus sign. Then read the relevant section below.

With Venus in Aries

This person enjoys the company of someone who is free-spirited and playful. While they may be attracted by your passionate intensity, they

There is a strong attraction here, but as Taurus and Scorpio are opposite signs there is likely to be some conflict. Like you, this person values stability and close attachment in relationships, but they have no interest in self-denial or sacrifice and they may find your emotional intensity to be hard work. What they *really* enjoy is comfort and physical pleasure. This person is very sensual and loves things of physical and material beauty. If you *do* connect it will be very powerful, and you may find the best place to do this is in bed – or on the couch maybe – or perhaps the kitchen table.

With Venus in Gemini

The Venus in Gemini person enjoys being light and chatty – not necessarily your strong point. They are much more detached in their relationships than you are, and they like to have a variety of friends to keep them amused. If you want to win this person's affection, don't try to tie them down too much or they are likely to become bored and restless. If they really like you, they will help you lighten up and be more detached. In return, you could open their eyes to a more primitive side of life.

With Venus in Cancer

This is a strong, harmonious contact. The Venus in Cancer person is able to tune in to your moods and respond with affectionate understanding. If you're in a rotten mood they will know what to do to make things better. They'll also understand when it's best to leave you alone. When your mood is confident and relaxed, the feelings between you will be easy and pleasurable. Like you, this person enjoys a close, private contact. If you feel safe together, you will be able to establish a very sensitive, caring relationship. This can be a very seductive combination.

With Venus in Leo

This connection can be strong, but Leo and Scorpio are quite different signs so be prepared to change your approach. The Venus in Leo person likes someone who is playful and outgoing. They may be quite self-centred, but if you pay them enough attention they will respond warmly. When you are moody and withdrawn they may try to draw you out in a friendly, playful way. On the other hand, they might throw a tantrum or simply lose interest. Both Leo and Scorpio are fixed signs so someone has to give a little for this one to work. When it does, the contact can be very passionate.

With Venus in Virgo

This pair is naturally complementary. Your serious, focused approach should meet with a helpful, practical response. Although this person may be critical, they offer their criticism in a friendly spirit. They are attentive to your needs, and if you have a problem they will be happy to help you sort it out. In relationships, this person is practical, caring and discreet, with a fine eye for detail (remember this when you are making the bed). If you are close, the connection is strongly emotional and physical.

With Venus in Libra

This person enjoys the company of someone who is refined, articulate and socially aware. Like you, they are strongly attracted to other people, but the contact they enjoy is detached and intellectual so your intense emotion may make them back off a little. Often they like to maintain a polite distance, so you will need to mind your manners if you want them to enjoy your company. If this person likes you, they will encourage you to be more detached and communicative. In return, you might persuade them to be a little less civilized!

With Venus in Scorpio

together, the contact will be deeply emotional and powerfully primitive. Perhaps you shared a cave in a previous life.

With Venus in Sagittarius

The Venus in Sagittarius person enjoys the company of someone with a sense of humour and a spirit of adventure. Whether their interest is in sport, travel or philosophical discussion, their preferred style is positive and energetic. If you try to tie them down too much they won't enjoy it. If you get jealous they might just laugh at you. Perhaps you could offer them stability while respecting their need for freedom. In return they would introduce you to new and pleasurable experiences. On an intimate level this person can be very playful and lustful.

With Venus in Capricorn

This contact is quite a compatible one. The Venus in Capricorn person is serious about their relationships and should respond well to your thorough, committed approach. They are willing to work hard for the sake of the relationship, and will encourage you to be realistic and down-to-earth. This person is strong and dependable, and in love they can be extremely sensual. Take your time, let the relationship build, and you may find it develops into a very productive partnership.

With Venus in Aquarius

This connection can be very strong, but the energies are quite different so there is bound to be some friction. The Venus in Aquarius person enjoys the company of someone who is honest and up front. They may value your serious, committed approach, but they like to maintain a clear, detached mind. When the emotions become too powerful they are liable to step back, and if they get a whiff of jealousy or possessiveness they're unlikely to stick around. This person is a bit on the eccentric side and enjoys a variety of social contact. You would do well to cultivate an air of detachment here. If you want to seduce them, start with interesting conversation.

With Venus in Pisces

This is a strong, easy contact. If this person likes you, they will be sensitive to your moods and will help you relax when you are feeling uptight. They will value your deep, emotional power, and may encourage you to direct it in a way which is helpful to others. When the mood is right and you both feel relaxed, your contact can be very close and romantic. You will heighten each other's emotional sensitivity, and may also connect on a psychic level. This can be a very seductive combination.

Famous Personalities with Mars in Scorpio

Martin Luther
Rembrandt
George Washington
Renoir
Robert Louis Stevenson
Mahatma Gandhi
Albert Schweitzer
Lawrence of Arabia
Harpo Marx
Yul Brynner
Rock Hudson
Robert Kennedy
Grace Kelly
Vanessa Redgrave
Jimi Hendrix
Danny de Vito
Pamela Stephenson
Grace Jones
Princess Anne
Mel Gibson
Olivia Newton-John
Susan Sarandon

Frank Zappa
Colette
Angie Dickinson
Stephanie Powers
Larry Flynt
Bruce Lee

9
Mars in Sagittarius

The difficult is done immediately. The impossible takes a little longer.

ANON

Sagittarius is the sign of the traveller and philosopher. With Mars in this sign, you have a passion for adventure. You want to be at the forefront of all activities, conquering new ground and making exciting discoveries. Being where the action is – this is what stirs your blood and arouses your passion. Naturally you have a strong desire to be free – to do what you like, when you like. There is a whole world out there waiting to be explored, and you don't want to be restricted by rules, regulations, authority figures or other boring, mundane limitations such as lack of money.

When you are fired up and things are going your way you will be warm and enthusiastic, looking to the future with hope and optimism and encouraging others to join you on your journey. In this mood you are a fun person with a great sense of humour.

On the other hand, if your desires are thwarted you can become annoyed and irritable, seeking to lay the blame on others while you search for a new way out. With Mars in Sagittarius, your forward-looking idealism needs to be balanced by practical understanding, otherwise you may be forever moving on to some future goal with an exaggerated sense of what might be and an unwillingness to take responsibility for the current situation.

Your desire for discovery and adventure could well lead to travel in far-off lands, meeting new people and exploring different cultures. But even if you don't travel overseas, you will still seek to broaden your [illegible] for new ideas may lead

[illegible]

Rudyard Kipling and Jack Kerouac – this was based on their own experience in foreign lands. On the other hand, Lewis Carroll's *Alice in Wonderland* was drawn purely from the realms of his own imagination.

You may also have a passion for sport, particularly the outdoor and adventurous sort. This will provide a healthy outlet for your powerful energies.

You have a strong sexual drive which you may express in a free and lustful way. In fact, you could be quite outrageous – like Oscar Wilde, whose sexual exploits led to one of the most sensational trials in literary history. On the other hand, your approach may be more strict and moralistic. In either case, your sex life will reflect your philosophical outlook. Ideally it will bring new experience and be a source of inspiration in your life.

The Woman with Mars in Sagittarius

Mars in Sagittarius is a powerful masculine energy – free-spirited, adventurous and highly assertive. As a woman, you may choose to tone this down a bit, but underneath, the passion remains. When this passion is aroused you will express yourself openly and honestly. You have a strong sense of what is right and wrong, and when called upon to give an opinion you can do this quite forcefully. Philosophical discussion will always arouse your interest, whether you participate with great seriousness or in a light-hearted, humorous way. Essentially you are inspired by the journey of life, and if you actively pursue this journey you will encounter many fascinating people and discover new and exotic places.

You are strongly attracted to a man who is open, honest and passionate – someone who lifts your spirits and gives you a sense of hope and optimism for the future. He will have a fun-loving side to his nature, but he can also be very serious – especially when important principles are at stake. This man wants to scale new heights and make new discoveries, and he will inspire you to share his journey. You could learn a lot from him – about yourself and about life – and share new pleasures along the way.

The Man with Mars in Sagittarius

You assert yourself most effectively by adopting a philosophical or moral standpoint based on strongly held principles. You may take a serious approach, arguing strongly for what you believe in and backing this up with weighty philosophical viewpoints. On the other hand, you might choose a more light-hearted approach, using humour to make your points. When you are feeling switched on and things are going well, you can carry the day with your warm, infectious energy. In this mood you will be generous and expansive, and other people will feel inspired by your optimistic attitude.

Your Compatibility with the Twelve Venus Signs

How will that special person respond to your Sagittarian approach? Before reading on, turn to the back of the book and look up his or her Venus sign. Then read the relevant section below.

With Venus in Aries

This is a strong, easy contact, and a very passionate one. The Venus in Aries person should respond readily to your open, free-spirited energy,

This person enjoys the company of someone who is practical and reliable. They have a strong appreciation of physical pleasure and beauty, and once settled into a comfortable rut are disinclined to move. They won't be impressed by your enthusiastic idealism unless they can see it as leading to something real and enduring. If you want to win this person's affection, you will need to discipline your restless energy and adopt a more down-to-earth approach. If this person really likes you, they will help you to be more stable and productive, and may also introduce you to some new physical and sensual pleasures.

With Venus in Gemini

This contact can be very strong, but Gemini and Sagittarius are opposite signs so there is likely to be some friction. This person will value your high, restless energy level, but will challenge your ideas by asking you to explain yourself and give logical reasons. They will enjoy sharing conversation, but they are more detached and their focus is most definitely in the here and now. When this contact works it will be highly stimulating, and you will bounce off each other with humorous chit-chat or serious debate. The sexual connection here can be quite potent.

With Venus in Cancer

The Venus in Cancer person takes pleasure in forming close emotional attachments. If you want them to like you, you will need to show that you are sensitive to their moods and attentive to their needs – to some extent at least. If you are loud and over the top, they are more likely to withdraw than respond with affection. A certain subtlety is required when wooing this person. Tread carefully, and give them time to learn to trust you. If and when they do, you will find them very caring and protective.

With Venus in Leo

This is a strong, easy contact. The Venus in Leo person will value your playful exuberance, and as long as you pay them enough attention they will respond with equal warmth and passion. This is a high-spirited connection, with plenty of fun and laughter to be had. When things are going well, you will really inspire each other. Join a sporting club, or perhaps an amateur theatrical society. You may become the star attraction on your social circuit. On an intimate level, the connection is very passionate and lustful.

With Venus in Virgo

This contact can be quite strong, but Virgo and Sagittarius are very different signs, so don't expect it to be easy. This person is likely to respond to your passionate, expansive approach with down-to-earth criticism. Be prepared to stop and think, and don't present your plan or idea until you have worked out the practical details. Even then they may still be critical, but if they really like you their criticism will be friendly, and they will be prepared to work hard to help you achieve your goals. If you are looking for intimacy, remember this person can be very fussy about cleanliness.

With Venus in Libra

This person will be attracted by your honesty and your desire to do the right thing, but you will need to cultivate tact and diplomacy. Don't expect them to respond with enthusiasm – they are more likely to be cool and detached. However, this doesn't mean they are not interested. Ideally they will have a calming effect, and help you keep things in perspective. Once you have established a good understanding, they will be more inclined to respond to your warm, passionate energy. Essentially this is a complementary connection.

you through thick and thin. You will also find they have a potent sexual nature and can be extremely seductive.

With Venus in Sagittarius

This could be a very powerful, passionate connection. If your minds are on the same wavelength, you will really be able to let go with this person. Together you can discover new places and explore new experiences, broadening your minds and having a ball along the way – the mile-high club perhaps? You may play sport or travel together. Sometimes you'll just sit around for hours talking. Unbridled lust is another definite possibility. The connection is warm, high-spirited and inspiring, and the only problem you are likely to have is knowing where to stop.

With Venus in Capricorn

In relationships, this person is down-to-earth and judgemental, and they won't be impressed by your big ideas unless they can see them leading somewhere practical. On the other hand, if you have thought things through and your goals are realistic they will be prepared to work hard to help you get where you are going. In love they are strong and dependable, but before you gain the benefits of their affection you will have to pass certain tests. Being worldly wise or having a large bank account helps enormously.

With Venus in Aquarius

This is quite a compatible pairing. If your values and principles are similar, you will find this is a good person to share your thoughts with. Their cool-headed response to your plans and ideas will be friendly and encouraging, and will help you maintain a clear, detached perspective. They enjoy socializing, and should be able to introduce you to some interesting new people. This could be a very liberating relationship between two people who are open to new ideas and experiences.

With Venus in Pisces

This person likes someone who is sensitive to moods and feelings. To win their affection you will need to reign in your high-powered energy and take a more gentle approach. If this person really likes you they will be very caring and supportive. When you feel down they will offer sympathetic understanding, and when things are going well they will enrich your relationship with pleasurable fantasy. In return you could bring fun and warmth into their life. This is a challenging contact, but when it works it is very stimulating.

Famous Personalities with Mars in Sagittarius

J.S. Bach
Lewis Carroll
Mark Twain
Oscar Wilde
Rudyard Kipling
Carl Jung
Greta Garbo
Judy Garland
Julie Andrews
Janis Joplin
Prince Charles
Patrick Swayze
Richard Nixon
John Travolta
Steffi Graf
Jack Nicholson
Warren Beatty
Joan Baez
Geena Davis
Faye Dunaway
Patty Hearst
Carmen Miranda
Patricia Neal
Tracey Ullman
Lesley Anne Down
Dustin Hoffman

Jack Kerouac
T.S. Eliot
Luciano Pavarotti

Lulu
Neil Diamond
Placido Domingo

10
Mars in Capricorn

Slow but sure wins the race.

PROVERB

With Mars in Capricorn, you have a passion for order and control. Perhaps the word 'passion' is misleading here – getting worked up into a wild frenzy is not your style, because your desire for control begins with your own passionate energy. You want to be in charge, first of yourself, and then of the whole situation, so your preferred approach is always cool-headed and self-disciplined. Sir Noel Coward is a good example. He was nicknamed 'Master', and was said to conduct himself with impeccable dignity.

You also possess great patience and determination. When you really want something you are prepared to keep working for a long time, moving gradually and testing every step like a sure-footed mountain goat as you climb towards your chosen peak. You're a natural builder, starting with a solid foundation and working your way up.

Your businesslike approach could take you far. You have a strong desire to attain material security and a stable position in society. You approach tasks in a carefully-structured way and are willing to take on responsibility when required.

John Wayne had these Capricorn qualities. He was a hard working, down-to-earth actor – always well prepared. Gregory Peck described him as having 'a saltiness and earthiness'.

Being a practical person, you are prepared to work within the rules and show due respect for authority because this way you get where you want to go. Ultimately though, you want to become the authority

to express this in a cold, hard, authoritarian way.

Your sex drive is strong, but you can keep it under wraps if you so choose – for a long time if necessary. Again, self-control is the source of your power. If you feel secure with someone, and are sure you won't be rejected, you can let yourself go and get right into it. Sex for you is very earthy and physical, and being in control turns you on. Sharon Stone, who has Mars in Capricorn, acted this out powerfully in the movie *Basic Instinct*. It's hard for someone to reject you when they are tied to the bed.

The Woman with Mars in Capricorn

There is a very businesslike side to your nature – cool, well-bred and very much in command. You want to wear the pants *and* the boots *and* the pin-striped suit. You won't make a big fuss about it, but you have the ability to take control and make sure things get done – your way. When you really want something, or someone, you will take a very practical, well-organized approach.

You are attracted to a man who has strength and self-assurance. He is solid and dependable, like the rock of Gibraltar. You are not impressed by the faint-hearted, sensitive type, or someone who makes a lot of noise but can't back it up in practice. The man you desire knows who he is and where he's going. He will more than likely have a well-bred air, and is perhaps a bit on the traditional side. If he doesn't have a good, solid background he will probably be in the process of creating one right before your eyes. This man has ambition, and is not afraid of hard work.

The Man with Mars in Capricorn

You assert yourself with great strength and determination, but you are not in any rush and you won't make a big show of it. Once you have worked out what you want and how you are going to get there, you have the ability to quietly take control and make it happen. Whether at work or in relationships, your approach is practical and systematic. Remember the fable about the hare and the tortoise? Well you are the tortoise. Before you take offence, remember, the tortoise won.

Nothing gives you a kick like being in control does. You are a realist, not an idealist, and you are looking for tangible results. You are not prepared to fight for something unless you can foresee a real, practical outcome. Even then, you probably wouldn't resort to fighting – it's such a crass way of getting things done. Usually you will look for a more sensible solution – one which just happens to work out in your favour.

Your Compatibility with the Twelve Venus Signs

How will that special person respond to your Capricorn approach? Before reading on, turn to the back of the book and look up his or her Venus sign. Then read the relevant section below.

With Venus in Aries

The Venus in Aries person enjoys the company of someone who is playful and spontaneous. If you want this person to like you, you had better hang up your suit and be prepared to let your hair down. While some people will find your mature, practical approach to be attractive, this person is more likely to think you are a bit stuffy. They are like a big kid looking for fun and adventure. Don't tell them about your latest successful business deal – take them bowling instead. Maybe you'll

tionship, if the Venus in Taurus person likes you they will respond by making things more comfortable, more beautiful and more pleasurable for you both. Whether this is business or personal, the two of you could create something very real and lasting. Just remember to pamper this person – a lot.

With Venus in Gemini

The Venus in Gemini person enjoys the company of someone who is light, detached and sociable. You will need to be a bit less serious and more outgoing if you want this person to like you. If you must talk business, try to make it amusing and informative. Lighten up, chat, play a few games – Monopoly, perhaps – then you will find them more responsive when you want to take charge and direct the proceedings. If you want to woo this person, find out where their intellectual interests lie, and make sure you have something interesting to say.

With Venus in Cancer

This connection can be strong and stimulating, but as Cancer and Capricorn are opposite signs there is likely to be some conflict. This person enjoys the company of someone who is sensitive to their feelings. By contrast, your approach is hard and practical. If you want them to respond, you will need to make an effort to tune in to their mood. In return, they may find your earthy strength to be helpful when their feelings are getting them down. If you really trust each other and feel safe together, you may find that opposites attract. When it comes to sex, this can be quite a potent combination.

With Venus in Leo

The Venus in Leo person will value your dignified, self-assured expression of power, but they may think you are a bit stuffy. If they really like you, they will try to draw you out of yourself in a warm and playful way. This person enjoys exercising power in relationships, so be prepared to hand them the reins sometimes. If you can do this graciously, and treat them with the respect they need, they will bring pleasure and laughter into your life. If you are close, they will also bring lust and passion.

With Venus in Virgo

This is a strong, earthy contact. The Venus in Virgo person will appreciate your practical, well-organized approach. As a friend, they will offer you useful, critical feedback which should help you achieve better results. When you are confident and take the initiative, they will fit in easily, adding their own touch of style in a precise, carefully thought out way. On an intimate level, this connection is very physical and sensual, though it may take you a while to get around to it – there are just so many things to organize.

With Venus in Libra

This can be a very strong, stimulating contact, but it's also challenging. The Venus in Libra person enjoys the company of someone who is gentle, refined and diplomatic. They like discussing things and negotiating a fair compromise, so at times they will find your approach too controlling and authoritarian. If you want this person to like you, you will need to listen to their opinion and spend some time coming to an agreement before you act. On a practical level, they could help you become more politically astute in your dealings with others.

With Venus in Scorpio

This contact is fairly strong and quite complementary. The Venus in Scorpio person values someone who is serious and committed in a relationship. They are extremely loyal, with very deep feelings, and if they trust you enough they will encourage you into a close, powerful, emotional liaison. Sex is very important to this person, and they can be

[illegible]

person could inspire you to move in new directions. If you are able to loosen up and have fun, you might discover some interesting new ideas and places together. On an intimate level, this person can be very passionate and playful.

With Venus in Capricorn

This is a very powerful, easy contact. The Venus in Capricorn person will value your solid strength and practical approach, and if they like you they will enjoy working to help you achieve your goals. In relationships, they are reliable and also very enterprising. Whatever you choose to do together you should find that your energies blend easily. Obviously this is an excellent contact for business. On the personal level, the connection is very earthy and physical – if you are not working too hard to get around to it.

With Venus in Aquarius

In relationships, this person is quite detached and enjoys conversation. Their ideas might be conservative or radical, or perhaps a curious mixture of both. In any case, they like to feel free to say what they think, and appreciate people who do the same. They enjoy socializing, and could introduce you to a variety of interesting new people. Depending on your intellectual compatibility, you may find this person fascinating or irritating. If you are interested, be honest with them – and don't expect them to follow orders.

With Venus in Pisces

Even though Pisces and Capricorn are quite different signs, they can complement each other well. This person will encourage you to be more sympathetic and open to feelings. If they really like you, they will be very attentive to your needs, and in return you could help them keep their feet on the ground. But don't be too hard and practical because this person is quite sensitive. They can also be very imaginative, and may bring a sense of romance into your life. If the contact works, you will find you balance each other's energies nicely.

Famous Personalities with Mars in Capricorn

Jane Austen
Franz Liszt
Leo Tolstoy
Albert Einstein
Noel Coward
John Wayne
Kirk Douglas
Marlon Brando
Doris Day
Jean Harlow
Woody Allen
Judy Collins
Harvey Keitel
Boris Becker
Jerry Seinfeld
Tom Selleck
Sharon Stone
Gerard Depardieu
Julia Roberts
Heather Locklear
Drew Barrymore
Shannen Doherty
Mia Farrow
Katharine Hepburn
Eartha Kitt
Margaret Mead
Roseanne
Loretta Swit
David Bowie
Sir Laurence Olivier
Kahlil Gibran
Bob Marley
Lee Majors
Patti Smith
Steven Spielberg
Rod Stewart
Brad Pitt
Prince Andrew

11
Mars in Aquarius

Since you weren't born a genius, then be a cantankerous fanatic.

ANON

You have a passion for thinking about life and communicating ideas. Possessed of a mind which is strong and active, you are very much the serious intellectual with a message. When your passions are aroused you are an excellent debater, arguing your case in a way which is strong, logical and convincing. You may become very attached to a particular point of view and hold it strongly, but you are also capable of changing your mind and you can do this quite suddenly and abruptly. Even though you may express your ideas with passion, you will speak most effectively when you remain detached – vehement, but not emotionally involved.

If your interest in something, or someone, is aroused, you can be quite an original thinker. When you *really* feel daring, your ideas may be downright outrageous. In fact, you get quite a kick out of shocking other people, don't you? But although you may be unconventional, you are also very rational and logical. Perhaps you will start with an unusual idea, then argue your case with such clarity and reason that others will find it hard to dispute your point of view. Or sometimes you will take several very reasonable ideas and weave them together in such a way that they lead straight to an original and unlooked-for conclusion. Just be aware, before you embark on your passionate and eccentric mental journey, of the silent conservatives in your audience.

They may not argue, but they will judge, and if they have the power to make your life difficult then you've got problems.

Behind your unconventional approach is simply a desire to be honest – for example, Lauren Bacall described herself as 'a girl who was dedicated to the truth'. You want to cut through the games and say exactly what's on your mind. But the truth is rarely fashionable, so your simple honesty turns out to be not so simple after all. Whether you intend to be radical or not, this is how you will appear to others when you speak your mind freely.

Your strong desire to communicate makes you a sociable person. Being part of a group turns you on, and you are able to align your energy with others and work hard towards a common goal. This may be something casual like a social event with friends, or perhaps you are involved with a more formal group of people who have ideas in common. This group could be associated with work or a hobby, or it might be a group fighting for a cause about which you are passionate. Many people with Mars in Aquarius develop a strong social conscience, and may fight on behalf of others for humanitarian ideals such as freedom, justice, or the rights of the individual. This was the case with Lauren Bacall, who became deeply involved in campaigning for freedom of expression in the movie industry during the late 1940s. She wrote, 'There's nothing like the charge you get out of being one of a group of people doing the same thing for the same reason – pure in thought and purpose, on a crusade. It's a fever.'

Your sex drive is erratic. You may be celibate for quite a while, then become extremely active. Talking about sex will always interest you – an example is Hugh Hefner, editor of *Playboy* magazine. Other famous people who have spoken their mind freely on this topic include Mae West ('I used to be snow white but I drifted.') and Robert Burns (author of *Nine Inch Will Please A Lady*). When you do get down to it, you are open to experimentation – 'I've tried several varieties of sex,' said Tallulah Bankhead. 'The conventional position makes me claustrophobic.'

The Woman with Mars in Aquarius

You are a strong-minded person, capable of communicating ideas

sharp as yours is. This person is quite eccentric in his own way. He has his own ideas about life and is not afraid to speak openly. He is also quite detached. In fact, sometimes he is so detached that you wonder whether he is from another planet. Once he sets his mind to something, nothing else exists, including you. But that's alright – you like your space, and the last thing you want is someone hanging around and clinging to you all the time. This man is different and intriguing. He's also an excellent friend – loyal, honest and sincere, the sort of person you can talk to about anything.

The Man with Mars in Aquarius

You assert yourself most effectively when you are clear and detached, certain of what you want to say, and able to back this up with well-reasoned argument. Your style is honest, friendly, sincere and original. You have no hesitation in putting forward an unconventional point of view, and as long as your logic is sound you will probably win the day. Once you get going, you can be quite a character. Some may think you're a bit crazy, but you know that this is the price you pay for having progressive ideas. You are just ahead of your time. Maybe one day the rest of the world will catch up.

Your Compatibility with the Twelve Venus Signs

How will that special person respond to your Aquarian approach? Before reading on, turn to the back of the book and look up his or her Venus sign. Then read the relevant section below.

With Venus in Aries

This is quite a complementary pair. The Venus in Aries person will value your desire for independence and freedom – they don't like to feel stifled in relationships. On the other hand, they enjoy the company of someone who is passionate and playful so they will encourage you to be less detached, to get into whatever's happening right now and have some fun. Loosen up, be spontaneous. This could be an enjoyable and feisty contact. On an intimate level this person can be very lustful.

With Venus in Taurus

This can be a stimulating connection, but Taurus and Aquarius are very different signs so you will probably clash a bit. The Venus in Taurus person likes a relationship which is stable and comfortable. Even though you may offer stability, this person is unlikely to feel comforted by your detached, intellectual approach. If you want to win their affection, then talk good sense. Be down-to-earth and practical, and make sure you attend to their physical needs – if they are not comfortable they'll have no interest in your clever talk no matter how well-intentioned it is. If this person really likes you, they will help you keep your feet on the ground. They could also introduce you to some new physical and sensual pleasures.

With Venus in Gemini

This is a strong, harmonious contact. Your compatibility begins on the intellectual level, and if you have shared interests you will bounce off each other with stimulating and pleasurable conversation. This person is not as fixed as you are, so they will encourage you to take a different perspective by throwing in a variety of new ideas. If they really like you, they will tell you clearly how they feel, and your connection will be warm and friendly. On an intimate level, talking about sex will add to the pleasure, and when you are apart your relationship could blos-

likes you, they will be very attentive to your needs, and also sympathetic and supportive when you are feeling down. In love they are extremely warm and affectionate.

With Venus in Leo

The contact here is very strong, but as Leo and Aquarius are opposite signs there is likely to be friction. This person should value your straightforward approach, and is likely to reply with equal candour. The question is, will you fight, or have a stimulating and pleasurable exchange? This is a warm person who relates spontaneously – from the heart, not from the head. As long as you don't expect them to agree with you all the time, you could have a lot of fun together. Just make sure you tell them how wonderful they are on a regular basis. On an intimate level, this person can be very lustful and playful.

With Venus in Virgo

Like you, the Venus in Virgo person is serious about relationships, and also quite intellectual. However, their mind is very practical, and they will interrupt to criticize if your ideas are too airy-fairy. Make sure your feet are planted firmly on the ground before you start talking. If this person likes you, you will find them caring and down-to-earth, with a fine eye for practical detail. When it works, this can be an interesting and stimulating contact between two quite different minds.

With Venus in Libra

This is a highly compatible connection. The Venus in Libra person enjoys the company of someone who has a clear, rational mind and knows how to communicate. They also like people who have a strong sense of fairness and justice. This person will encourage you to look at both sides of an argument before committing yourself to a particular point of view. Their gentle, diplomatic response will help you soften your stance and gain a more balanced perspective. This is a warm, friendly contact, and there will always be plenty to talk about. If you are close, your conversations will continue on a more intimate level in the bedroom.

With Venus in Scorpio

The connection here can be very strong, but not necessarily easy. In relationships, the Venus in Scorpio person is very private and deeply emotional. They need to really trust someone before they will respond with affection. If you are too detached, they will be wary of you and won't open up. You need to tell this person about your deeper feelings. If you do, and they respond, you will find out what deep emotional attachment is all about. This person has a strong sexual nature and can be very seductive.

With Venus in Sagittarius

This is a very complementary pair. The Venus in Sagittarius person enjoys the company of someone who has vision and ideas. Your eccentric ways will probably make them laugh, and they will encourage you to be more passionate and spontaneous. Together you could enjoy exploring new people, new places and new ideas. This could be a high-energy connection which is liberating for both of you. If you share a common vision, the contact will be particularly close. If you share a bed, it will be particularly lustful.

With Venus in Capricorn

In relationships, the Venus in Capricorn person is hard-working and down-to-earth. If they really like you, they will encourage you to put your ideas into practice and work towards material success. On the other hand, if they think your ideas are impractical they will tell you so in no uncertain terms. In a close relationship this person is strong and

ate your desire for independence. Like you, this person needs space in their relationships. They enjoy having a variety of friends, and don't like being tied to a jealous partner. Together you could form a very strong bond of friendship which still allows you to move freely in your respective social circles. This could be a very liberating and pleasurable relationship, both sexually and intellectually.

With Venus in Pisces

This person enjoys the company of someone who is sensitive to moods and feelings. If you are too hard and detached in your approach, they won't like it. On the other hand you may share a sense of idealism, and if you show a genuine concern for other people the Venus in Pisces person is likely to respond with approval. You need to relax and create a softer mood for this person – maybe then you can drift away together to a more pleasurable place. If they really like you, they will be very devoted and caring.

Famous Personalities with Mars in Aquarius

Robert Burns
Mae West
Leonardo Da Vinci
Cary Grant
Howard Hughes

Lorne Greene
Lauren Bacall
Charlton Heston
Hugh Hefner
Robert Wagner

Joanne Woodward
Edward Kennedy
Grace Slick
Peter Cook
John Cleese
Pete Sampras
Linda Lovelace
Tallulah Bankhead
Queen Elizabeth II
Farrah Fawcett
Bridget Fonda
Jane Fonda
Mary Queen of Scots
Winona Ryder
Zsa Zsa Gabor
Ryan O'Neal
Marcello Mastroianni
Tennessee Williams
Truman Capote

12
Mars in Pisces

Reality...has a sliding floor.

RALPH WALDO EMERSON

With Mars in Pisces, you have the most direct, self-centred, assertive planet in a sign which is very fluid and impressionable. You are strongly affected by the prevailing mood, drawn along by a tide of feeling. Is this tide going in your direction? Did you choose it? Do you want it, or is someone else pulling your strings?

The fact is, you have a strong desire to feel part of something greater. Your feelings are very sensitive, and you are able to tune in to other people, pick up on their vibes, and direct your energy in sympathy with theirs. You are a very impressionable and highly imaginative person, but you need to strike the right balance between asserting yourself clearly and strongly and maintaining your easy-going sensitivity.

Because you are concerned with keeping the peace, you may find yourself going along with other people in an effort to avoid conflict. In this frame of mind you may also try to get your own way by manipulating others, rather than risking a more direct, open approach which might cause trouble. Alternatively, you may seek to escape the situation altogether. Sometimes these methods offer the perfect solution, but in large doses they will create confusion and misunderstanding, and you may get a reputation for being unreliable and untrustworthy.

This imaginative and impressionable side to your nature is both a strength and a weakness. Vincent Van Gogh and Billie Holiday are examples of people who became victims of their own sensitivity. On the other hand, they seduced and enchanted millions of people with their artistry. Errol Flynn and Marilyn Monroe did the same. Possibly the greatest seducer of all time was Casanova. What a shame he didn't make any movies.

When you are clear about what you want, confident you can achieve it, and feeling relaxed, you have a special ability to pick up on the current mood or fashion and direct it with your creative imagination. When others are moving in the same direction, you can align your energy with theirs and carry them along with your enthusiasm.

Ultimately you will fulfil your deepest desires when your chosen path is one which also benefits other people. You have a strong urge to be part of something which goes beyond narrow self-interest. Maybe you want to help people who are outcast and suffering, or perhaps you will bring people together through involvement in the arts or business. Elizabeth Taylor does both of these. Whatever your domain, you have the ability to tune into other people's feelings and inspire them to work together for a common cause. Like the conductor of an orchestra, you can work with many different players at once, uniting them in the creation of something quite enchanting. In business you make a great strategist, though you may prefer to leave the practical details to someone else.

On the personal level, you are sensitive and sympathetic, and can be a source of inspiration to others. Whether you are conscious of it or not, you also have a strong psychic ability.

Your sex drive is powerfully emotional. You are very seductive, and you have a strong yearning for a contact which will transport you to another world – 'the mystique of the adventure' as Casanova put it. The mood is all-important, and to create that special atmosphere you may use music, candles, perfume, alcohol or some other stimulant. One of Casanova's specialities was the oyster orgy. Errol Flynn tried the A to Z of aphrodisiacs – some of them quite exotic such as one containing crushed pearls and fine specks of gold. He had mixed results. In the end he declared: 'There is only one aphrodisiac – the special woman you love to touch and see and smell and crush.'

The Woman with Mars in Pisces

[illegible] and a desire to experience the [illegible]

[illegible] person who yearns for something greater. [illegible] escape with, perhaps through sex or music, or watching lots of movies together. He is definitely sensitive. Perhaps he is an artist or a spiritual visionary, or he could be just a soft, gentle person. You relate strongly to his idealism, and you have the power to comfort, help and inspire him to rise above himself. Your emotional bond will be very close, and when you share an ideal you are able to work well together, combining your creative imaginations to produce something quite special.

The Man with Mars in Pisces

When you are feeling relaxed and confident you have the ability to assert yourself strongly while retaining your softness and sensitivity. Other people will respond to your relaxed, easy-going approach, trusting that you have their interests at heart as well as your own. Your energy has a calming effect, and you are able to influence others with your sympathetic and considerate approach. More than this, you have the ability to deal with many different people and situations at the same time, drawing them together through your creative focus and inspiring them with your imagination. Your vision is your strength, and your sensitivity to others will help you enlist their support in achieving your goals. You are at your best when you strike the right balance between satisfying your own desires and helping other people achieve theirs.

Your Compatibility with the Twelve Venus Signs

How will that special person respond to your Piscean approach? Before reading on, turn to the back of the book and look up his or her Venus sign. Then read the relevant section below.

With Venus in Aries

The Venus in Aries person likes people who are high-spirited and playful. They have a sense of mischief, and their need for pleasure is self-centred and immediate. They won't respond enthusiastically if you are too vague and spaced out. Be bold and different. Shock them with outrageous comments, take them to places they've never been to. This person is more interested in adventure than candlelit dinners – unless, of course, they've never had a candlelit dinner. On an intimate level they can be very lustful.

With Venus in Taurus

This connection can be quite complementary. The Venus in Taurus person will encourage you to direct your imaginative energy into the realm of physical beauty and sensual pleasure – a visit to an art gallery, or dinner and the theatre. And afterwards, satin sheets. This person enjoys their physical comforts. They value people who are down-to-earth, and can be quite judgemental if they think you are being impractical. But if they really like you, this judgement will be given with affection and should help you keep your feet on the ground.

With Venus in Gemini

This contact can be strong, but Pisces and Gemini are quite different in nature so there is likely to be some friction. The Venus in Gemini person enjoys the company of someone who is detached, rational and communicative. If you are being vague and dreamy, and not making sense, they are likely to become irritable and impatient. On the other hand, if this person really likes you they will help you put your thoughts and ideas into words. This can be a stimulating, fast-moving contact between two active imaginations. Just make sure you are thinking clearly before you approach this person.

will be sympathetic and protective,
mutual seduction is an ever-present possibility.

With Venus in Leo

The Venus in Leo person enjoys the company of someone who is confident, playful and outgoing. They like to bathe in their partner's glory, so if you aren't feeling too glorious, it's best if you stay away – you are unlikely to find sympathy here. However, when you are on top and feeling good this person will enjoy your company. If they really like you, they will draw you out of yourself in a warm, friendly and playful way. Just remember to shower them with compliments on a regular basis. If you want to seduce them, do it with style and flair.

With Venus in Virgo

Pisces and Virgo are opposite signs, so although this contact can be strong and complementary it may also bring conflict. The Venus in Virgo person likes someone who is practical, discriminating and well-organized. These are not your obvious strengths, but still, this person could be attracted to your sensitive feelings and easy-going nature as a foil to their hard-headed practicality. They will challenge you to keep your feet on the ground, but if you can handle the criticism it will help you turn your dreams into reality. In return, you could help them relax and get in touch with their feelings, bringing romance into their life. When it comes to sex, this can be quite a potent contact.

With Venus in Libra

This person will value your sensitivity, but would like you to express yourself clearly and rationally. For you, this is not always easy to do, but if you want to win this person's affection you need to make the effort. Their sensitivity is more detached and intellectual than yours, and they enjoy the company of someone who is refined, balanced and tactful in their approach. If they really like you, they will encourage you to be more detached and communicative. In return, you could help them get more in touch with their feelings. If you want to seduce this person, begin with interesting conversation.

With Venus in Scorpio

This is a strongly emotional and extremely harmonious contact. If they trust you enough, the Venus in Scorpio person will respond to your advances with great warmth and passion – and I mean *Passion*. This person's sexual appetite is legendary among astrologers. In relationships they are loyal, protective and deeply affectionate. When you are feeling lost and out of sorts, they will help you regain your focus and feel confident again. This is someone you can really depend on, even when things get rough. If you are close, your connection will be very seductive.

With Venus in Sagittarius

The Venus in Sagittarius person enjoys the company of someone who is honest, friendly and outgoing. They also like someone who has clear moral principles and acts accordingly. You need to be up front with this person. If you are evasive or deceptive they will turn off very quickly. On the other hand, if you get on well you may find you share a dream, in which case they will give you plenty of positive feedback and friendly encouragement as you work together towards your goal. Once they've checked you out and decided you are okay, this person can be very playful and lustful.

With Venus in Capricorn

This contact can be quite complementary. In relationships, the Venus in Capricorn person is hard-working and dependable. As long as they think your goals are achievable they will stick by you. When they think you are being unrealistic they will become judgemental, but if they really like you their criticism will be offered with affection and you

[illegible]

tion when it obscures logi[illegible]

is clear before you approach them. If this person likes you they will encourage you to be more detached and independent. In return, you might persuade them to give their brain a rest and let go with a bit of feeling. If you share a concern for broader social issues this will add another dimension to what could be a most interesting relationship.

With Venus in Pisces

This is an extremely powerful contact – very emotional and romantic. This person should respond readily to your relaxed, easy-going approach. If they really like you they will tune in to your feelings and follow your moods with great sensitivity. When you need to escape from the hard, cold realities of life, they will be sympathetic and understanding and will know what to do to take your troubles away. Whether at work or in your personal life, together you could create something which is enchanting and inspiring. Perhaps you share a strong compassion for other people and can do charitable work together, or maybe you are both into the arts. On the personal level you could enjoy many pleasurable hours exploring an endless variety of ways to seduce each other.

Famous Personalities with Mars in Pisces

Michaelangelo
Casanova
Vincent Van Gogh
Spencer Tracy
Errol Flynn
Billie Holiday
Lana Turner
Marilyn Monroe
Shirley Temple
Johnny Cash
Elizabeth Taylor
Bob Dylan
Michelle Pfeiffer
Ralph Nader
Debbie Reynolds
Ava Gardner
David Janssen
Allen Ginsberg
Simone De Beauvoir
Glenn Close
Martina Navratilova
Camille Paglia
Lisa Marie Presley
Gertrude Stein
Tina Turner
Alan Alda
Elton John
Burt Reynolds
Annie Lennox
Björn Borg
Bo Derek
Carrie Fisher
Steve McQueen
Billy Crystal
Linda Hamilton
Tom Hanks

PART 2

What's Your Pleasure?

13
Venus in Aries

To be alive is to be burning.

N. BROWN

In love and friendship you are like a refreshing, innocent child – passionate, playful, and fun to be with. You express your affection openly and honestly and you appreciate people who respond with equal directness.

This childlike character is the essence of the beauty of many actors with Venus in Aries – for example Audrey Hepburn, Shirley Temple, Liza Minnelli and Marilyn Monroe. All of these film stars have different styles, but the innocent beauty of Venus in Aries is something they each project to great effect.

Aries is naturally a naive and impulsive sign which thrives on spontaneity. With Venus in Aries, you will enjoy life most when you are living for the moment, allowing your feelings to come tumbling out without too much calculation or self-consciousness. Obviously this is not the most tactful or thoughtful way of operating, but if you have to stop and think about the other person too much it kills the spontaneity, your energy level drops and the pleasure starts to disappear. This is why you are self-centred in relationships – being true to yourself is the best way you know to have fun. However, although you may be self-centred, you are generally able to express this in such a charming way that others still find you attractive. Or should that read, you are generally able to express this in such a charming way that you get away with it?

Your enthusiasm and natural sense of adventure makes you good at initiating friendships. You love the excitement of meeting new people [illegible] talent for getting parties off to a lively start.

[illegible]

a degree of independence. [illegible] always be on the look-out for new ways of sharing passion and pleasure with your partner.

You love to play games, and your sense of fun can be quite mischievous at times – even aggressive in a friendly way. For you, relationships need to have spark and passion otherwise there is no pleasure. So, to keep the excitement going, you might play tricks on someone, or stir them into a playful fight. Friendly competition of any sort will bring enjoyment. If you are the intellectual type, you may like playing mind games or board games. When you are feeling good you have the ability to make a game out of any situation. Given the right mood, you could even turn a shopping trip into an entertaining competition. Marilyn Monroe particularly enjoyed teasing. She would drop outrageous comments in a casual way and wait for the reaction. Her innocent manner was part of the trick, and she used it to full effect.

If you are more the physical type, you will take great pleasure from sport. Whether you play, or whether you prefer to be an enthusiastic spectator, being caught up in the energy and emotion is the source of your enjoyment. The excitement of competition may attract you to sport, but there is more to it than that. Physical exercise also keeps your body fit and strong, and the image of youthful potency is something you value highly. In your mind, beauty and potency go hand in hand.

With all this passionate energy coursing through your veins it's a short step from love to lust. As Isadora Duncan described it, 'I sat back and looked at him, and suddenly felt my whole being going up in flames like a pile of lighted straw.' At the right time, all it takes is a small gesture and you will follow your passion with enthusiasm. Obviously this may land you in trouble but, as the Spanish dancer said in the movie *Strictly Ballroom*,'A life lived in fear is a life half lived.' For Venus in Aries this could read 'A life lived in fear is a life half loved.' Other factors in your chart may modify this side of your nature, but you will always find that the most exhilarating love involves an element of risk. As with all games, sometimes you win and sometimes you lose. Your next move may be an embarrassing blunder, or it could propel you into the affair of a lifetime, but you will never know if you don't take the chance.

The Woman with Venus in Aries

With Venus in Aries, your beauty lies in your fiery personality. When you feel good you are warm and outgoing, with a mischievous sense of fun and a passionate love of life. Shyness does not become you. Wear bright colours and dress in a way which highlights your strength, whether that is physical strength (you might like to dress up in sports gear) or the strength of your personality.

When you feel good, you are high-spirited and independent in all your relationships, and not afraid to say what you think. If other people don't like it, that's their problem. If they can't handle you, they'd better go and find someone more tame and obedient. In love you are playful and feisty. You don't mind a bit of conflict as long as it is done in a friendly spirit. In fact, if there is no conflict there is no excitement, and then life becomes very dull and flat. When things are going well, your energy is strong and sparkling, and you are a delight to be with.

The Man with Venus in Aries

Although Venus is a feminine planet, Aries is a strongly masculine sign, so it should be easy for you to express this side of your nature.

You see relationships as a challenge, and approach them with a sense of adventure. Meeting someone new is always exciting. After the initial [illegible] worn off you may go looking for someone else, but if

[illegible]

kids sharing an adventure. [illegible]

ways, and you certainly won't agree about everything, but disagreement is part of the fun – don't call it a fight, call it sport and enjoy it. You wouldn't swap her for a bevy of obedient women – well, perhaps for a while, just to see what it's like.

Your Compatibility with the Twelve Mars Signs

How will your Venus in Aries personality respond to the other person's approach? Before reading on, turn to the back of the book and look up his or her Mars sign. Then read the relevant section below.

With Mars in Aries

The chemistry generated by this contact can be powerfully electric – instant passion. I guess you could have a platonic relationship with this person, but that would be a bit of a waste, so let's hope this is not someone who is out of reach. The only couple I know with this combination was a fairly staid pair of intellectuals who scandalized their work colleagues by falling suddenly in love and running off together. Given the right circumstances – or even given the wrong circumstances (which makes it more wicked and therefore more exciting) – this could be a dynamite relationship. Whether you engage in heated conversation, play vigorous sport or head straight for the bedroom, the contact is pure and strong.

With Mars in Taurus

This person's approach is down-to-earth and determined. They may take a while to make up their mind, but once they decide they really want something, or someone, they are not easily put off. While you value their strength and determination, at times you may see them as being too preoccupied with practical and material matters. If you can be patient and put some of your energy into helping them out, then you might cajole them into loosening up and having some fun. Once this person gets their mind off work and on to pleasure they can be very sensual.

With Mars in Gemini

This person has a passion for communicating ideas which you should find very stimulating. Once they get going, they can be very witty and entertaining, and if your minds are on a similar wavelength your contact will be strong and energetic. You could enjoy playing intellectual or physical games together. Whether these games are on the sporting field, in the lounge or in the bedroom, this person will make sure there is plenty of variety, and your spirited response should keep things moving at a lively pace. This could be a high-energy contact with plenty of fun and laughter.

With Mars in Cancer

This connection can be very stimulating, but it is also challenging. The Mars in Cancer person expresses their feelings strongly, and becomes very attached to home and family. Emotional security is very important to them. If you are really interested in this person, then sometimes you will need to button your lip and toe the line, especially when they are in a bad mood. They may feel threatened by your need for independence, so you will have to reassure them at times. Again, it depends on their mood – and don't expect them to tell you – you need to learn to read their weather report. If you are still interested, you will be glad to hear that when this person is feeling good, they are quite capable of matching your lustful energy.

With Mars in Leo

This contact is strong, easy and very upbeat. The Mars in Leo person wants to play, and the only condition is that you tell them how won-

With Mars in Virgo

One thing Aries and Virgo have in common is a need for independence, but there the similarities end, so if you want to attract this person's attention you will have to work at it. When this person gets fired up about something, or someone, they become very serious and analytical, and if they don't think the situation is perfect – which they invariably never do – they will also be highly critical. You'd better take them seriously, for a while at least. If their critical analysis starts to inhibit your sex life, perhaps your best approach would be to agree that there is a problem and suggest an extended experiment to fix things up. For example, you might spend several months systematically trying every position in the *Kama Sutra*.

With Mars in Libra

Relationships are very important to this person, but they will approach them in one of two different ways. On the one hand they might be quite aggressive, in which case you will clash – but perhaps in a pleasurable and exciting way. On the other hand they may be extremely polite and friendly, and then you will probably feel a mischievous urge to have a go at them. While you express your feelings with great passion, this person tries to remain cool and detached – even when they are angry. In this case, opposites attract, and while you encourage them to warm up and let go they will challenge you to think about things in a more detached, rational way. When this contact works it is very stimulating, and may provide a good balance between sex and interesting conversation.

With Mars in Scorpio

This person has a deep and secret passion. You can feel it simmering below the surface, but the question is how are you going to draw it out? Tread carefully, give them time to trust you, and show a bit of loyalty. If you succeed, you will find that under their carefully controlled surface lies enough energy to fuel a volcano, so once you pass the security checks and get down to it, there is an inexhaustible well of lust and passion to enjoy. Just be aware of what you are unleashing, and don't take it too lightly. If you change your mind, extricating yourself could be a complex and tricky business.

With Mars in Sagittarius

The contact here is playful and passionate. When this person is feeling confident, you will find yourself strongly attracted to their warm, outgoing energy. They are never short of ideas, and always seem to be going somewhere interesting. This relationship can be a real adventure – sometimes it will be serious and enlightening, while at other times you will simply have fun together. Your shared interests may be in sport, the arts, travel, or exploring the spiritual dimension of life. As long as things don't become too predictable, there should be a high sexual energy between you.

With Mars in Capricorn

This connection can be very stimulating, but there is likely to be friction. While you like relationships which allow you a certain amount of freedom and independence, the Mars in Capricorn person wants to be in control. You may be prepared to compromise and follow their rules to some extent, but if they try to restrict you too much you are likely to rebel. If you are close, you will probably be able to do this in a friendly way – maybe you can help them laugh at themselves a bit. In return, this person can bring stability to the relationship. It's very much a matter of give and take here. On the personal level, the contact is earthy and passionate.

[illegible]

neous, they will [illegible] can be a very warm friendship between two very independent people. It could also be wickedly enjoyable.

With Mars in Pisces

This person wants to escape and take you with them. If they really know where they are going, and you like the idea, you could have a great time together, but when they are feeling depressed or confused you will find their lack of clear direction to be exasperating. If you really like this person, you will learn to adjust to their moods. When they are feeling down you may be able to jolly them out of it in a friendly, affectionate way. On the other hand, when this person feels confident they can be very charming and seductive, and together you could create a highly pleasurable relationship of lust and romance.

Famous Personalities with Venus in Aries

Michaelangelo
Abraham Lincoln
Charlotte Brontë
Queen Victoria
Tchaikovsky
Renoir
Sigmund Freud
Isadora Duncan
Albert Einstein
Henry Fonda
Jean Harlow
Perry Como
Lana Turner
Marilyn Monroe
Shirley Temple
Audrey Hepburn

Edward Kennedy
Johnny Cash
Elizabeth Taylor
Rudolf Nureyev
Judy Collins
Liza Minnelli
Harvey Keitel
Jack Nicholson
David Janssen
Doris Day
Mia Farrow
Katharine Hepburn

Jayne Mansfield
Elizabeth Montgomery
Lesley Anne Down
Peter Fonda
Elton John
Sir Laurence Olivier
Roy Orbison
Bob Marley
Steve McQueen
Pete Townsend
Kurt Russell
Tony Blair

14
Venus in Taurus

It is better to have a permanent income than to be fascinating.

OSCAR WILDE

With Venus in Taurus you have a very earthy, sensual side to your nature. Your appreciation of physical pleasure is strong, and whether your favourite indulgence is cooking, gardening or art, you enjoy sharing this with the one you love.

An extensive wardrobe of fine clothing is another pleasure commonly enjoyed by those with Venus in Taurus. Princess Diana had a wardrobe the size of ten terrace houses. When choosing a white blouse, she had to decide between velvet, silk, satin, cotton and lace. If only.

Your love of physical beauty and pleasure extends naturally into luxury. Let's start with your body. It needs to be comfortable, right? Not too hot, not too cold, and pleasantly cushioned in your favourite chair or in the arms of your doting partner who knows *exactly* what you need. Now that you are comfortably settled, what can we pop into your mouth? Nature has such rich and bountiful fruits, and there are some you *really* love. Perhaps some juicy, succulent grapes, or a piece of that delicious strawberry cheesecake you couldn't resist buying at the supermarket this morning? On the other hand, you might have grown the strawberries yourself and spent a pleasant afternoon in the kitchen whipping up a batch of delicious goodies. Now, there is something missing – is it music, or the cat curled up on your feet, or a massage perhaps? Yes, that's it – a massage. And while your partner is rubbing you with fragrant oil you can occupy yourself usefully by going through the luxury hotel brochures and planning your next weekend retreat. Who said over-indulgence was a fault? You've raised it to an art form.

With such a strong appreciation of physical pleasure and comfort, a large bank account comes in handy. There may be some things money can't buy, but they're not at the top of your shopping list. When you feel rotten, buying things will cheer you up – a friend of mine calls it retail therapy. When you feel good, shopping will make you feel even better – and of course, you will enjoy it more if you do it with a friend.

You tend to form relationships which are strong and stable. Once you find someone with whom you can share a comfortable rut, you will not be inclined to move. Not that you are anti-social, but who wants to risk an uncomfortable evening of forced conversation and disappointing entertainment when the alternative is a guarantee of pleasure with the one you love in the comfort of your own home? Not you.

The down side to forming such strong attachments, whether to an object or a person, is you find it difficult to let go. In fact, once you get used to something, or someone, you can be downright possessive. You have such an ability to feel physically close to another person that you may end up seeing them as an extension of your own body – the extra arm that puts the kettle on or the pair of feet that walk to the shops to fetch lunch.

In love you are down-to-earth and reliable. You express affection in a very physical, sensual way – through touch and massage, or by giving things of material value and beauty. During her affair with Gary Cooper, who had Venus in Taurus, the actor Patricia Neal said 'At least a couple of times a week Gary would arrive in the early evening with flowers or wine, sometimes a small bag of groceries or little things for the apartment.' Of all the earth signs, Taurus is the most luxuriously indulgent.

The Woman with Venus in Taurus

Your expression of beauty is very physical and material. You are par-

extend into the garden, where you could take pleasure in growing things you find rich and beautiful.

The world of the arts will also attract you, especially when it appeals directly to the senses – a rich melody, a beautiful painting, or perhaps a fine piece of sculpture. You may just adorn your life with such beautiful things, but on the other hand you may actively express yourself through the arts. To some extent it depends on other factors in your chart. One thing is certain, your attractiveness as a person is strongly tied to your active appreciation of the good things in life, and sharing this with others will bring you much pleasure.

The Man with Venus in Taurus

You enjoy having a comfortable routine in your life which revolves around familiar pleasures. Your appreciation of physical beauty and comfort is strong, and once you get used to something you like you will happily return to it over and over again. You may not realize it, but you have the ability to develop your appreciation of physical beauty to a very high level, and also to create things of great richness and value. You could do this with plants in the garden, with food in the kitchen, or through art.

You idealize a woman who can bring a predictable routine of comfort and pleasure into your life. This woman is very earthy and sensual.

She is a superb cook, and has a fine appreciation of all things rich and beautiful. In a relationship she offers you stability, so she is more likely to be conservative than radical. Most of all she knows how to pamper you and fill your senses with pleasure.

Your Compatibility with the Twelve Mars Signs

How will your Venus in Taurus personality respond to the other person's approach? Before reading on, turn to the back of the book and look up his or her Mars sign. Then read the relevant section below.

With Mars in Aries

This person has a wilful side to their nature, and a strong sense of adventure. If you are interested, be prepared to leave behind your comfortable routine – for a while at least. Eventually you may be able to coax them into something more cosy and predictable, but don't expect to ever tame them completely – they are just too independently minded. If the relationship lasts, this person will always challenge you to take risks. They could bring passion and excitement into your life, and in return you could make sure you travel together in comfort and style.

With Mars in Taurus

The attraction here is very powerful and very physical. This person has a passion for rich, sensual experience, so you should be able to introduce them to some earthy delights which will be well received. They also want stability, and are prepared to work hard to achieve it. If you settle into a comfortable routine, the relationship could be long and productive, with the potential for ongoing pleasure only limited by your combined bank accounts. When it comes to sex, this is a very potent contact.

With Mars in Gemini

This person has strong, restless, intellectual energy, with a passion for conversation. If you are open to it, they will introduce you to a whole variety of new ideas and social contacts. This is not what you would call a comfortable connection, but it's certainly stimulating. If you really like this person, they will inspire you to add to your repertoire of

to some good old-fashioned food therapy. When this person feels confident they can be extremely seductive. There is definite potential here for mouth-watering combinations of food, sex, and other pleasurable indulgences in the privacy of your own home. For ideas and inspiration, try watching the movie *Like Water for Chocolate*.

With Mars in Leo

This connection can be very stimulating, but there is likely to be some conflict. Once this person gets fired up they are very passionate and outgoing, and just as fixed in their desires as you are attached to your favourite pleasures. If you lock horns, they will try to dominate you with the power of their personality while you become negative and judgemental. On the other hand, if you allow yourself to be charmed by this person they could bring a great deal of fun and laughter into your life. In return, you might enrich theirs with comfort and beauty. This can be a great combination for throwing a party – they create an upbeat and entertaining atmosphere, while you attend to the physical comfort and sensual enjoyment of the guests.

With Mars in Virgo

The contact here is strong and harmonious. This person's approach is very practical and well-organized, so your down-to-earth response will fit very nicely. While they create an environment which functions efficiently, you can enrich it with some of the good things in life. If anything breaks down they will make sure it is fixed, and if you make a bit of a mess their passion for order and tidiness will ensure it is cleaned up. They may inspire you to be more discerning and meticulous in your work, while you could encourage them to appreciate the richness of physical beauty and pleasure. On an intimate level, the connection is very earthy and physical.

With Mars in Libra

This person has a strong desire to communicate, and they express their passion in a detached and rational way. Like you, they are attracted to beauty, but their approach is refined and intellectual rather than earthy and sensual. Perhaps they could entertain you with poetry or conversation while you whip up a gourmet meal. Whatever your pleasure, they will encourage you to develop a more refined taste, paying particular attention to balance and harmony. They will also want to talk about things, so it is important to share your thoughts with this person – stimulating discussion turns them on.

With Mars in Scorpio

This is a strong connection, and a very stimulating one, but as Taurus and Scorpio are opposite signs there is bound to be some conflict. The Mars in Scorpio person will express their passion with great power and feeling. It won't always show on the surface, but underneath that calm exterior is a very determined person. Even though they may threaten to disrupt your comfortable world, you will probably find yourself drawn to this person with a fascination for what you might discover. They may be very demanding, even ruthless, but if you can handle this and strike a balance between their needs and your own, the contact will be highly stimulating. Once drawn together it is difficult to separate you.

With Mars in Sagittarius

When this person feels confident, their energy is expansive and outgoing. They have a desire to explore new ideas and travel to new places. If you want to attract them, you will need to cultivate a more exotic taste. Perhaps you could study up on different cultures – their art, their cuisine, their sexual habits. Then you could tour the world together, sampling it at your leisure. This person wants to get out and have fun.

[illegible]

tune selling art. The Mars in Capricorn person has plenty of initiative which is nicely complemented by your earthy stability. On the personal level the contact is strongly physical, though you will have to wait until this person has completed their work and got things in proper order before they will want to indulge themselves.

With Mars in Aquarius

This contact can be very stimulating, but it is also challenging. The Mars in Aquarius person expresses their passion in a detached, intellectual way. Once they get fired up they will be so busy talking they probably won't notice that the fire's out and the dinner has gone cold. If you want to attract this person, then you'd better cultivate a detached interest in their favourite topics of conversation – sex is likely to be one of these. If you really like them, your interest will be genuine, and they will challenge you to think seriously about things while at the same time amusing you with their eccentric ways. In return, you could help them keep their feet on the ground and encourage them to develop their taste for physical pleasure and beauty.

With Mars in Pisces

This person expresses their passion with a softness and sensitivity which complements your down-to-earth style in relationships. They can be very imaginative, and also extremely seductive, with the ability to create an atmosphere which is quite enchanting. Your fine appreciation of comfort and physical beauty should fit in well with this, and together you could create a relationship with the perfect balance between mysterious romance and pleasurable predictability.

Famous Personalities with Venus in Taurus

Leonardo da Vinci
Oliver Cromwell
Thomas Hardy
Charlie Chaplin
Adolf Hitler
Spencer Tracy
Gary Cooper
Salvador Dali
John Wayne
Joan Crawford
Bette Davis
Marlon Brando
Doris Day
Paul McCartney
Hayley Mills
Grace Jones
Michael J. Fox
Prince Phillip
Billy Joel
Princess Diana
Cyndi Lauper
Steffi Graf
Jessica Lange
Debbie Reynolds
Allen Ginsberg
Glenda Jackson
Imelda Marcos
Simone Signoret
Suzi Quatro
Herb Alpert
Warren Beatty
Ryan O'Neal
James Taylor
Tennessee Williams
Prince Edward

15

Venus in Gemini

Interruptions are the spice of life.

DON HEROLD

For you, the essence of pleasure is a variety of light entertainment and social contact to stimulate and amuse your mind. Enjoyable conversation is near the top of your list. Whether in person or over the phone, you take great pleasure in talking with others and you like to keep the conversation moving through a variety of different subjects.

Your tone is light and detached – being tied down in a long, heavy exchange is not your idea of a good time. Tony Curtis, for example, has been nicknamed 'Mr Anecdote'. Brief, amusing stories, jokes and informative chit-chat come naturally to you, and if the other person can join in you will take pleasure in their company.

Being such a friendly, sociable type, you will naturally develop a mind which is alert, witty and informed. Then you can charm others with your skilful self-expression. This is the essence of your beauty, and you like to be told that you're a fun, interesting person to be with.

Given your natural curiosity and your quick, active mind, reading will provide another source of pleasure – whether it's newspapers, magazines or books, they keep you informed and ensure you are never short of something new to talk about. Uma Thurman, who has Venus in Gemini, is a big reader with a large collection of books. Games and puzzles will also arouse your curiosity, and sharing these with friends provides yet another way of enjoying yourself.

You may also choose to develop your appreciation of the arts. Depending on other factors in your chart, you may or may not develop artistic skills yourself, but you will certainly appreciate the skills of others. Perhaps you have an interesting collection of music, a variety of beautiful clothes, or a love of poetry or the theatre. Whatever your taste in art, there will be plenty of variety, and you will enjoy sharing your appreciation of it all with friends and loved ones.

In relationships you are versatile and adaptable, with a low boredom threshold and a strong curiosity for new experiences. Obviously you will enjoy a variety of relationships. Henry VIII of England certainly did. Most people with Venus in Gemini employ a less extreme method of ensuring this. While some may be downright fickle, most are just in the process of educating themselves. As Uma Thurman said, 'You learn as much as you can from every relationship – most aren't meant to last a lifetime – and you keep looking for the one.'

Your natural air of detachment in relationships gives you the freedom to explore without losing yourself in passion. John F. Kennedy had many affairs, but never fell head over heels. 'I'm not the tragic lover type,' he said.

For you, love and friendship need to provide an endless variety of things which are pleasing to the mind. Essentially, Gemini is a rational, intellectual sign, but you don't need to be top of the class to express yourself in a clever, witty and eloquent way. You just have to value your own mind and take pleasure in communicating.

The Woman with Venus in Gemini

As a woman with Venus in Gemini, you love conversation and enjoy socializing. You will always be happy to put aside your work when a friend phones for a chat. You may also be an avid reader. Games and puzzles will also interest your active mind. You have a strong intellectual curiosity which needs stimulation, preferably from a variety of sources. At your social best, you can charm, amuse and fascinate at least half a dozen people at the same time. You will never be short of something interesting to say – some little witticism or a brief, amusing story. This is your beauty – you are an eloquent, social charmer, capable of captivating an audience with your witty, active mind.

[illegible]

different friends you have. But although they [illegible] has the ability to amuse and entertain you. For any one woman to really captivate you, she would need to be witty and sociable with a number of interesting sides to her nature. Together you may share an interest in the arts, playing games of all descriptions, or simply doing the social rounds. Venus in Gemini is light and airy. This doesn't mean that you won't have a deeper side to your nature, but in love and friendship your preference is for light entertainment and detached conversation.

Your Compatibility with the Twelve Mars Signs

How will your Venus in Gemini personality respond to the other person's approach? Before reading on, turn to the back of the book and look up his or her Mars sign. Then read the relevant section below.

With Mars in Aries

This is a moderately strong, easy connection. If your minds are on the same wavelength, this person could inspire you with their enthusiasm and new ideas. Their passionate approach will stimulate you in a pleasurable way, and your detached, cool-headed response will provide balance and variety. If your shared interest is in sport or travel, this person has enough initiative to get both of you going. This is a cheeky, witty, energetic contact and you should complement each other nicely in play, whether at the card table, on the tennis court or in bed.

With Mars in Taurus

This person takes a practical, no-nonsense approach and likes to call the shots. Once they decide they want something, or someone, they show great determination in pursuing their goal. If you really like this person, make an effort to be dependable. If they think you're playing around, they are liable to become extremely jealous and impossible to live with. You may need to explain things from your point of view, just to reassure them. This person is capable of hard work and commitment, offering you stability and security. In return, you can help them to lighten up, communicate and enjoy themselves.

With Mars in Gemini

This is an extremely powerful contact. The person with Mars in Gemini has a strong, active mind which you should find very stimulating. They will be able to come up with an endless variety of ideas to keep you amused and entertained. Games, tricks, puzzles, jokes, clever one-liners, amusing stories – and all delivered with such passion! This person can really captivate you with their clever mind and social skills, and the speed with which they move around will ensure that you are never bored, whether at a restaurant or in the bedroom.

With Mars in Cancer

This person is strongly emotional, and it's important for them to know where they stand. Don't play games with their feelings. Try to tune in to their moods, and take them seriously. When they feel good you can be your normal detached self, but when they are down, stay with them and let them know you care. If you are close, this person could transport you to a world of rich and powerful emotion and in return you could help them chase away their bad moods with friendly humour.

With Mars in Leo

This is a strong, easy contact. When the Mars in Leo person is feeling good about themselves, they can entertain you with their warm, enthusiastic personality. Their positive energy should stimulate your mind pleasantly, and you will find it easy to be with them. In return, the things you say will be generally well-received by this person. Just remember to keep the ego-boosting comments flowing when they're feeling down. This could be a very sociable contact between two people who stimulate and

may simply feel that whatever you say

slow down and think more seriously about what you say, and the Mars in Virgo person can lighten up a little, then you may be able to compromise between pleasant discussion and heavy analysis. In this case you will find your difference in style is very stimulating.

With Mars in Libra

This is a strong, easy contact. Essentially it is a meeting of minds. You will find the Mars in Libra person is very passionate about ideas – this may be personal stuff or their ideas could be about broader social issues. You may also share an interest in the arts. While this person initiates conversation, your replies will add interest and variety, and you should have no trouble in keeping each other entertained till the early hours. Whether you do more than talk is open to discussion.

With Mars in Scorpio

This person is deeply passionate and emotional, but they won't always show it openly. You need to tune in to the way they are feeling and respond with sensitivity. Sometimes your light-hearted chatter will draw them out of themselves in a pleasant way, but when they are not in the mood, forget it. If you are close, this person will provide you with a stable focus while you lighten up their life with humour and interesting conversation. Just keep in mind their capacity for jealousy and revenge before you play around. I guess you have two choices – either don't do it or keep it a secret.

With Mars in Sagittarius

This is a very powerful connection. Despite being opposite in nature, the attraction is strong. Mars in Sagittarius is an inspirational sign with a passion for philosophical banter. Like you, they enjoy communicating, but their ideas may be so grand that they can't always express them in a clear, rational way – this is where you can help. When things are going well you will bounce off each other in a delightful game of conversation and laughter. Both of you enjoy tricks and jokes, though Sagittarius has an abstract bent while your cleverness is more logical. When this person is in the mood, they can be very playful and lustful.

With Mars in Capricorn

While you enjoy sharing light, intellectual pleasure, this person is much more serious, with a passion for work. However, if you value the work they do, you should be able to respond with a variety of clever ideas and social contacts which may contribute to their success. This is a down-to-earth, enterprising person who could offer you stability in a business or personal relationship. When the day's work is done, you could help them lighten up and have some fun.

With Mars in Aquarius

This is a strong, easy contact. The Mars in Aquarius person can be very passionate about ideas and causes, and you should find it easy to respond with a variety of interesting replies. They could inspire you to think more deeply about life, while you help them lighten up, distracting them with a bit of humour perhaps. Together you could enjoy detached, stimulating conversation about anything which grabs your active minds. This is a relationship which starts on an intellectual level. Whether it goes any further is something you could no doubt talk about for hours.

With Mars in Pisces

This is a powerful connection, but the energies are very different so it won't be straightforward. You may find this person to be frustrating because they can't always explain what they want. While you value someone who knows their mind and expresses themselves clearly, this person has an elusive and dreamy side to their nature. When you are [illegible] understand what they

[illegible]

- Henry [illegible]
- William Shakespeare
- Rembrandt
- Rudolph Valentino
- Gregory Peck
- John F. Kennedy
- Tony Curtis
- Jacqueline Kennedy Onassis
- Julie Christie
- Ringo Starr
- Bob Dylan
- Cher
- Brooke Shields
- Eva Peron
- [illegible]
- Andre [illegible]
- Gabriela Sabatini
- Emma Thompson
- Jerry Seinfeld
- Dudley Moore
- Isabelle Adjani
- Candice Bergen
- Olivia Hussey
- Isabella Rosellini
- Uma Thurman
- Joanna Lumley
- Tom Hanks
- Helen Hunt

16
Venus in Cancer

Contentment lies more often in cottages than palaces.

PROVERB

In love and friendship you are extremely sensitive, deeply emotional and very warm and caring. Because of your sensitivity you will be careful to protect yourself. You need to feel safe and secure with someone before you allow them to get too close, so although you may be warm and caring towards people generally, you reserve something special for the ones you really trust. These trusted friends are sensitive souls like yourself, and with them you share a feeling of closeness and intimacy.

Judy Garland was born with Venus rising in Cancer, and expressed great emotional sensitivity in all her relationships. 'Let me tell you, legends are all very well if you've got somebody around who loves you,' she said. A less obvious example is Ernest Hemingway, who was quite a macho man. But it was said that 'anyone close to him knew he was really soft and sentimental'.

Your strong feelings result in strong moods. To some extent these moods will reflect how those around you are feeling. Like a sponge, you soak up these feelings and vibrate in sympathy. But your moods can also indicate how you feel about yourself. If you feel insecure and unloved you may become over-sensitive and too easily hurt. In this state your moods will be crabby, and you may also become possessive and cling too hard in relationships. On the other hand, your sensitive feelings are also the source of your beauty and attractiveness. When you feel good about yourself they are very strong and nourishing for

[illegible]

to treat all your friends and loved ones [illegible] family unit. These feelings may extend even further to include an attachment to your local community. As the family base, your home is also important to you.

Being in love can bring out the psychic side of your nature quite strongly, and the closer you feel to those you love the stronger your psychic connection will be. This connection may extend to a place you love, whether this is your home, your home town or your home country. You feel a strong attachment to places which give you a sense of belonging.

This sense of belonging is further enhanced by exploring the history of the people and places you love. You may enjoy tracing your family tree or discovering the past life of your home town. More broadly, Venus in Cancer can show a love of history generally. Anything which enhances your feelings of security will bring you pleasure.

In love, your emotional sensitivity is heightened. You are kind and gentle, and also quite sentimental – sharing pleasurable memories with your partner will bring you closer. Your love is also very private, and this strengthens the bond of intimacy between you.

The Woman with Venus in Cancer

Your beauty as a woman lies in your kind, loving nature. You are ultra-sensitive, and can't help being touched by other people's feelings. Your natural inclination is to respond with care and sympathy, particularly towards those who are helpless and vulnerable. You may be like a mother figure to your friends, supporting them when they are down and helping them feel safe and secure. If someone you care for is threatened, you will be very protective.

Although you relate to people generally with care and sensitivity, the people you feel close to will have a special place in your heart. In love, you are very attentive to the other person's needs in both a practical and emotional way. You have a very delicate emotional antenna which will pick up on their changing moods and help you respond in sympathy. When you love, you do so deeply and faithfully, and the feelings you share are very private.

The Man with Venus in Cancer

With Venus in Cancer you have a very sensitive side to your nature which will come out in your relationships. If you feel uncomfortable showing your feelings too openly, you will tend to express them by doing practical things which show you care. When a friend comes to visit, you will be attentive to their needs, making sure they are comfortable and offering them drinks and snacks. If you know someone is feeling down you may phone them for a chat or pop in for a cup of tea. You pick up quickly on other people's moods, responding in sympathy. Naturally, you will enjoy the company of people who are themselves sensitive and caring.

You idealize a woman who has very sympathetic, motherly qualities. She is attentive to your needs and well tuned in to your moods. When you feel down she knows what to do to make things better. She will enjoy being around the home, making sure you are comfortable and that your needs are met. As well as being very sensitive and affectionate, she also knows your taste in food.

This woman is a gentle, thoughtful person, and also quite sentimental. Your bond of trust lies deep, and you are extremely protective of each other. The love you share is very private, and not something you will necessarily show to others.

It was her 'home-body qualities' that attracted Errol Flynn to his third wife, Patrice Wymore. He raved about her Kansas home-baked cooking, and commented 'No one ever believes I want some peace,

[illegible]

look up his or her [illegible]

With Mars in Aries

This contact may be very stimulating, but can also be challenging. When the Mars in Aries person is fired up they are very direct and assertive. You may feel the need to protect yourself, hiding your sensitive feelings in case you get hurt. This person wants passion and adventure. If you feel brave enough you may choose to take the risk and open up to them, in which case you are in for some excitement. At the same time you may help them get more in touch with their feelings. When this contact works it is very passionate and emotional.

With Mars in Taurus

This connection is quite complementary. It will probably start slowly, but once the Mars in Taurus person gets going you will find their energy is strong and very dependable. Together you could build a relationship which is stable and nourishing. The Mars in Taurus person will ensure that your practical, material needs are met, while you provide emotional care and support. This could be a very good home/family combination. If you are close, the contact will be warm and sensual, very physical and emotional.

With Mars in Gemini

When this person takes the lead they are detached, intellectual and highly communicative. If you want to maintain their interest you will need to put your feelings into words, preferably in an interesting and eloquent way. Make sure you are well informed on their favourite subjects, and then your conversation could be stimulating and pleasurable. If you get on well, this person will keep you entertained with their wit and cleverness. In return, you will encourage them to express their feelings, making the contact more intimate.

With Mars in Cancer

This is a very powerful and emotional contact. Like you, the Mars in Cancer person has very sensitive feelings, and they can express these with passionate intensity. The more you trust each other and open up, the closer you will be – and you really can get *very* close to this person. When they express themselves openly and you respond, your emotional connection will be so strong that the rest of the world may as well not exist. This can be a potent contact – emotionally and sexually. The key to achieving a good rapport lies in feeling safe and secure enough together to really relax and let go. If you do connect on this level, you will be extremely protective of each other.

With Mars in Leo

When this person feels confident, they will express themselves with great passion and enthusiasm. They may tend to boss you around, and could be a bit insensitive at times. Because of this you might feel a need to protect yourself, but if you get on well you will find they draw you out of your shell in a warm and friendly way. This person wants to have fun. As long as you show loyalty and respect, they will be generous and exciting to be with. In return, you will be supportive, bringing intimacy to the relationship. Sexually, this person can be very lustful.

With Mars in Virgo

The connection here is complementary. This person's approach is down-to-earth and discriminating. They may fuss and worry a bit, but they only do this because they want things to be just right. If you really like this person, you will help them settle down and relax. You could work well together around the home, creating an environment which is comfortable and well-organized. While they attend to the practical matters, you will ensure it is a place where people feel at home. This [illegible] be a very caring, thoughtful relationship.

[illegible]

you really like them and want to [illegible] detached and logical in your response. This person is not usually turned on by warm, fuzzy feelings, and if you become over-emotional they are unlikely to be impressed. When the contact *does* work, they will help you articulate your feelings, giving you a new perspective on yourself in relationships. In return, you can help them get out of their head and in touch with their feelings.

With Mars in Scorpio

The connection here is strong and harmonious. When this person is switched on they can captivate you with the intensity of their feelings. They won't go shouting it from the roof tops, but you'll pick it up at the gut level and, if you like them, you will find it easy to respond. This contact is pure emotion. You are both very sensitive and feel a need to protect yourselves, so you may be wary of each other to start off with. But if you decide to open up and trust each other, you will find you have a capacity for deep intimacy and perhaps also a psychic connection. This is a mutually seductive combination, and a very private one.

With Mars in Sagittarius

This person has a passion for adventure, both physically and philosophically. Whether they are discovering a new idea, a new job or a new relationship, they have strong principles, a sense of humour and tons of enthusiastic energy. You may feel bombarded by this person, and inclined to withdraw. On the other hand, they might stimulate you to take the risk and go on a journey with them. If you do, they will open your eyes to a whole new world – be it physically, intellectually or sexually. In return, you could offer them warmth and emotional support.

With Mars in Capricorn

This can be a very strong and stimulating contact. It may be a case of opposites attract, or you might find your differences create difficulties. Perhaps it will be a bit of both. This person's approach is practical and well-organized. They want to take control and get things done – their way. You may feel they are trying to steam-roller you, without consideration for your feelings. On the other hand you might appreciate their strong, down-to-earth approach as the perfect complement to your softness and sensitivity. The more you value your differences and work together, the better this relationship will be. If you are close, the contact will be physical and emotional, and very stimulating.

With Mars in Aquarius

This person's approach is detached and intellectual, and sometimes quite eccentric. They have a passion for communication, so if you want to hold their interest make sure you know their favourite topics of conversation and have something intelligent to say. You may find this person has a social conscience which matches your compassion. They will challenge you to take a different view of yourself and the world, while you can help them relax and get more in touch with their feelings.

With Mars in Pisces

This is a strong, harmonious contact. The Mars in Pisces person will approach you with an easy-going sensitivity to which you could readily respond. If you trust this person and feel safe with them, the feelings will flow readily between you. You will be very sensitive to each other's moods, treating one another with care and sympathy. This person is extremely seductive. If they are feeling confident and you are in the mood, intimacy should follow naturally and easily.

Famous Personalities with Venus in Cancer

Florence Nightingale

[illegible]

Jimmy Stewart
Errol Flynn
Dean Martin
Liberace
Yul Brynner
Judy Garland
Tom Jones
Meryl Streep
Princess Anne
Arnold Schwarzenegger

[illegible]
Marquis de Sade
Jerry Garcia
Carlos Santana
Björn Borg
Ray Davies
Peter O'Toole
Lindsay Wagner
Keanu Reeves
Elizabeth Hurley

17
Venus in Leo

Always star in your own movie.
KEN KESEY

In love and friendship you are warm, playful and high-spirited. You may be a bit self-centred, with a tendency to see your friend or partner as an extension of yourself, but when you are feeling good you will get away with it because your energy is so positive. You enjoy being the centre of attention, the life of the party, and when things get going you can help everyone else have a good time. Then if they choose to shower you with praise and compliments you won't mind a bit.

You possess a dramatic flair which comes out strongly when you are feeling good. Naturally you enjoy playing games which give you the opportunity to display this side of your nature – like charades, for example.

Leo relates to royalty, and in relationships you can be very noble and generous – like a king or queen. But like a king or queen, you also expect loyalty from your subjects, I mean friends. For you, respect is a very important part of love. If you feel belittled or rejected, and your pride is hurt, you will turn off very quickly – you may even throw a bit of a royal tantrum.

Like the regal person, you are very proud, and you may also be a bit vain at times. It may be true that before you can share love with others, first you must love yourself – but try not to overdo it. And while we're on the subject of your faults (no disrespect intended), you can also be quite jealous and possessive. These traits tend to go hand in hand with loyalty.

... company of someone who makes you

[illegible]

Kidman, both of whom have Venus in Leo ... admiration club with two members.

In love, you express yourself with great warmth and generosity. For you, making love should be playful and energetic, with plenty of fun and laughter. It should also be a pleasurable contact between two people who admire and respect each other. And of course, it *must* be special – like a royal performance with a cast and audience of two.

The Woman with Venus in Leo

Your beauty lies in your sparkling personality. When you feel good, you are warm, friendly and vibrant. In this mood you are fun to be with, and you will naturally become the centre of attention. Other people will be attracted by your good vibes, and will want to join in. You have the ability to make people laugh and feel good about themselves. This is a real gift.

You may enhance your personality by dressing to make an impact, so that people will notice you when you walk into a room. Although you relate to people with a certain childlike innocence, your style is also proud and very noble – like royalty. While you may enjoy playing games, as a friend you are true, loyal and generous towards those you love. Essentially you relate to others as a warm, upbeat person with style and flair.

The Man with Venus in Leo

Although Venus is a feminine planet, Leo is a very masculine energy, so you should have no trouble in expressing this side of your nature. When you feel good in a relationship you are loyal and sincere, and extremely generous. You love to play games and have fun, and you enjoy the company of people who can match your playful, outgoing energy.

You idealize a woman who has beauty and style, someone you could be proud of. When you enter a room together, heads will turn. Perhaps it's her clothes or make-up, or the way she does her hair, but more than likely it will be her personality – Leo is all about personality and making an impact. This woman is not just your partner – she's a star, and together you are a knock-out duo. She may be self-centred at times, and perhaps a little vain, but she is warm and honest, and also very loyal.

The artist Rubens had Venus in Leo. He was married twice, both times to beautiful models who he painted hundreds of times for all the world to admire. He was totally loyal to each wife, and apparently very happy – his last child was born nearly nine months after his death at the age of 62.

Your Compatibility with the Twelve Mars Signs

How will your Venus in Leo personality respond to the other person's approach? Before reading on, turn to the back of the book and look up his or her Mars sign. Then read the relevant section below.

With Mars in Aries

The connection here is strong and harmonious, with a high energy level. When this person feels confident they will inspire you with their direct, spontaneous approach. They have a passion for exploring new experiences and new people. You should find their enthusiasm very infectious, although they may not always treat you with the respect you so richly deserve. Still, this could be a very warm, exciting and playful relationship. It could also be extremely passionate and lustful.

[illegible]

stubborn, so if you clash there [illegible] maintain this person's interest, you will have to put up with a bit. However, if you can sort out your differences, this could be an extremely sensual and passionate relationship as well as a productive one.

With Mars in Gemini

This is quite a complementary connection. The Mars in Gemini person has a passion for communication, and when they are switched on their mind is quick, sharp and witty. Their high energy level will stimulate you, and together you can share plenty of fun, games and laughter. While they inspire you with their clever conversation and humour, your passionate response will encourage them to be more flamboyant. Socially you could be a very dynamic pair. There are also artistic possibilities here.

With Mars in Cancer

This person's approach is sensitive and very emotional. They are able to create an atmosphere which is warm and inviting, but it's important for them to feel safe before they really open up. If you like this person, try to tune in to how they are feeling, and respond in sympathy. Once they trust you, you will be able to take a few more risks, then your warm, infectious personality will encourage them to be more outgoing and playful. When they feel confident and secure, this person can be extremely seductive, with a combination of strength and sensitivity which is most attractive.

With Mars in Leo

This is a strong, passionate connection. If this person feels confident and assertive they will approach you in an open, warm-hearted way. They want to play, and can be quite theatrical. If you feel inspired to join them, you could have a lot of fun together. You may compete a bit for the centre of attention, but as long as you have a healthy respect for each other your games will be warm and friendly. Together you could co-star in your own movie – a sizzling tale of lust and passion perhaps?

With Mars in Virgo

This person's approach is very methodical and down-to-earth. They have a passion for analysing things, and a desire for perfection in everything they do. If you are attracted to them, be prepared for a bit of criticism and try not to take it the wrong way. It's not that they want to hurt you or bring you down, but achieving perfection requires serious thought and hard work. You may learn to tidy up your act a bit and improve your style in a practical way. In return, you could encourage this person to let go and enjoy themselves, bringing something special into their life.

With Mars in Libra

This is a complementary connection. The Mars in Libra person is very sociable and communicative, although their approach can be quite detached. They may be confrontational, or very gentle and polite. Either way, you will find they have a particular charm which inspires you to be more thoughtful and fair-minded. When they are becoming too serious and intellectual, you will encourage them to loosen up and have fun. If you share an interest in the arts this will be a source of great pleasure to you both. It's also a very good contact for socializing as a couple.

With Mars in Scorpio

This can be a very stimulating contact, but it's also a challenging one. The Mars in Scorpio person is deeply passionate, but they won't necessarily show it openly. They tend to be quite secretive about their desires and need time to develop trust before they open up. If you try to draw them out of themselves before they are ready, they are likely to become wary and retreat. If you really like this person, give them time to get to know you. Meanwhile, try to tune in to their moods and

enthusiasm. As long as you are on the same philosophical wavelength, you should find it easy to respond. They will inspire you to explore new ideas, people and places, and together you could share a great deal of pleasure and laughter. If you are looking for a relationship which opens up exciting new horizons, this could be just the right match for you. There may also be a strong lustful energy between you.

With Mars in Capricorn

This person takes the initiative in a very practical and businesslike way. They won't make a big show of it, but they want to be in control. When they are too domineering for your liking you will probably tell them so in no uncertain terms. If this doesn't faze them, they will simply try a different approach. This person has great patience, and they are able to persevere until they get what they want. If you really like them, they may inspire you to develop a more mature, professional style. In return, you will encourage them to be more flamboyant. 'Who wears the pants?' could be the eternal question.

With Mars in Aquarius

Aquarius is the opposite of Leo and the attraction is strong, but the energies are very different so there is likely to be some conflict. This person's approach is direct, but quite detached. They have a passion for honest communication, and can be quite blunt at times. Don't expect them to pander to your ego. They will tell you exactly what they think, and if you can't handle it they will probably lose interest. If you don't take offence, you will find them very exciting to be with. They will challenge you to look at yourself in a new way and to think about the way you relate to others. This level of self-honesty can be very liberating. In return, you could help them loosen up and have some fun. When it comes to sex, this can be a very potent contact.

With Mars in Pisces

This person's approach is relaxed and easy-going. They are sensitive and considerate, and when they really know what they want their charm is quite irresistible. However, their moods are changeable, and they won't always be as positive and outgoing as you would prefer. If you really like this person, learn to read their mood and adjust your response accordingly. When they feel down, try to be sympathetic. After a while, the weather will change and you will be able to draw them out with playful humour. If you are close, the contact will be passionate and full of feeling.

Famous Personalities with Venus in Leo

Leo Tolstoy
George Bernard Shaw
Alfred Hitchcock
Greta Garbo
Lauren Bacall
Charlton Heston
George Sand
Andy Warhol
Madonna
Michael Jackson
Sylvester Stallone
Tom Cruise
Olivia Newton-John
Pete Sampras
Belinda Carlisle
Coco Chanel
Phillis Diller
Gina Lollobrigida
Victoria Abril
Carole Lombard
Dalai Lama
Marcello Mastroianni
Barbara Cartland
Whitney Houston
Peter Finch
Truman Capote
Nicole Kidman
Linda Hamilton

18
Venus in Virgo

She is so industrious, when she has nothing
to do she sits and knits her brows.

ANON

In love and friendship you are caring and practical, with a strong sense of duty. You are very attentive to other people's needs, right down to the smallest detail.

You like to show concern about your friends' health and wellbeing, and will enjoy helping them improve themselves in any practical way. You take pleasure in seeing your relationships become more healthy and work better. To this end, you like giving constructive criticism. Unfortunately, other people won't always take it the way it is intended. Sometimes it is better to say nothing, and instead show your affection by doing something they will enjoy – like giving them a massage, fixing them something nice to eat or helping them with their shopping perhaps. Bob Hoskins said of Robert de Niro, who has Venus in Virgo, 'He's a real friend. He's helped me shop for my wife's and my kids' Christmas presents.' Another big star with a modest side is Heather Locklear. Friends and co-workers have described her as 'nice, sweet, down-to-earth'.

Because your sense of beauty is tied in with practicality, you may develop a particular talent for craftwork. Making something of beauty can obviously bring pleasure, but if it is also useful then so much the better.

You enjoy analysing your relationships and sharing this process with friends and those you love. The point of this exercise is to help create better relationships. For you, a good, loving contact should bring improvement to both individuals, as well as to the relationship itself. Obviously you will be attracted to people who are caring and practical like you, and who will work beside you to build a strong, healthy partnership.

On the other hand you may choose to live alone, if not for the whole of your life then at least for certain periods. Of all the signs, Virgo is the most modest and self-contained.

As a lover you can be very earthy and sensual, much more than your modest demeanour would suggest. But you are still fussy, particularly about cleanliness. This fussiness may inhibit you, making you fairly prudish and shy of letting go. But not necessarily – Virgo can go either way. Lawrence of Arabia kept to himself for his whole life, and was virtually sexless. By contrast, Pauline Bonaparte – sister of the famous emperor – was one of the most licentious women of the nineteenth century, with a particular liking for well-endowed men. Even so, she made a fetish of cleanliness, bathing every morning in a bath filled with hot water and twenty litres of milk. She was also attended regularly by a number of doctors.

The Woman with Venus in Virgo

Your beauty lies in your serious, caring nature, and your genuine concern for the wellbeing of those you love. You are an earth woman and a very practical one, but there is nothing crude or coarse about the way you express this. On the contrary, you are neat and meticulous with a fine eye for detail. In your dress and manner you are impeccable – flawless, if that's possible. Sophia Loren is a good example.

In love you possess a self-effacing shyness which is most attractive. You like to be of service to those you care for, and you carry out this work discreetly and efficiently. Although you may have an air of purity, Virgo is still an earth sign, and underneath that neat and tidy exterior is a very sensual woman.

The Man with Venus in Virgo

... sensible, down-to-earth person. You can

You idealize a woman ...
also very caring and faithful, and likes to make a fuss of you. If you have a problem, she will always be ready to sit down and discuss it with you, and you enjoy hearing what she has to say because you respect her fine, analytical mind. She will probably also be health-conscious, very tidy and well-organized in her personal habits. This is a modest woman who takes great pleasure in attending to your needs.

Your Compatibility with the Twelve Mars Signs

How will your Venus in Virgo personality respond to the other person's approach? Before reading on, turn to the back of the book and look up his or her Mars sign. Then read the relevant section below.

With Mars in Aries

This person's approach is direct and spontaneous. If they really want something, or someone, they usually come straight to the point. You may find their impulsive energy exciting and refreshing, or you might simply think they are rude and thoughtless. This person can be very energetic and enterprising, but they usually have a few rough edges. If you really like them, you will encourage them to be more practical and discriminating, and help them become more discerning in their approach. In return, they may inspire you to be a bit more bold and adventurous in love.

With Mars in Taurus

The contact here is strong and harmonious. When this person takes the lead they are very reliable and down-to-earth, and they don't give up easily. You will value their practical, no-nonsense approach, and should find it easy to respond with affection and support. If you have any constructive criticism, they will probably take this well because they are always interested in becoming more productive. Together you could build a relationship which is solid and enduring. On an intimate level, the contact is very physical and sensual.

With Mars in Gemini

The attraction here may be strong, but you are likely to have your differences. This person wants to demonstrate their skill and intelligence. They have a passion for communicating, and when they are fired up their mind is sharp and witty. You may value their cleverness, but their style is quite detached and they are not always very practical. You enjoy communication which is purposeful, that leads somewhere useful, so sometimes this person's chatter may seem shallow and pointless. In this case, you will become critical. However, when you are interested in what they have to say, they will inspire you with new information and ideas. In return, you could encourage them to direct their thoughts in a more practical and discerning way.

With Mars in Cancer

The connection here is quite complementary. This person expresses their passion in a very caring and deeply emotional way. Their combination of strength and sensitivity can be most attractive, and your practical, thoughtful response provides a good balance for their strong emotions. This could be a very caring and supportive relationship, and a very private one. You help them keep their feet on the ground, while they enrich your life with feeling. If you are feeling a bit shy and unsure of yourself, this person could draw you out in a seductive and pleasurable way.

With Mars in Leo

When this person feels confident and assertive, they are very warm and outgoing. They may try to impress you with their style and gen-

With Mars in Virgo

The connection here is very strong and harmonious. This person's approach is careful, methodical and down-to-earth. They have a passion for analysing, and are prepared to work hard at getting things just right. You should be able to work well with this person. Their practical approach and fine eye for detail will please you, and they will be encouraged by your serious, thoughtful response. When there is a problem, you should be able to talk it through and find a solution. Together you could create a relationship which is close enough to perfect to satisfy you both. There is a strong sexual energy between you, if and when you feel so inclined.

With Mars in Libra

This person has a strong desire to relate to others, and their style is rational and detached. At times they are very courteous, but they can also be quite argumentative. You will appreciate this person's refined, intellectual nature, and should enjoy talking to them, but your interests may differ somewhat. While you like analysing practical problems and looking for solutions, this person is more interested in social contact and sharing ideas. If you spend time with them, they could introduce you to new people, and help you develop a more sophisticated awareness of social issues. In return, you will encourage them to be more self-critical and down-to-earth. This can be a very thought-provoking contact.

With Mars in Scorpio

The connection here is quite complementary. This is a deeply emotional person who expresses their passion with great power and determination, but they don't make a big show of it. You will value the depth and thoroughness with which they work, and will feel quite comfortable with their desire for privacy. Your thoughtful, practical response should be received well by this person, and together you could get on quietly with whatever you wish to achieve. This can be a very private contact between two caring, devoted people. If you are close, you will find this person very seductive.

With Mars in Sagittarius

The connection here is stimulating but challenging. When this person is fired up they will approach you with great enthusiasm. Life is an adventure, and it turns them on. Usually they have a philosophical bent, and love discussing ideas and theories. If they have a plan, they will want to tell you all about it, but although their ideas may be inspiring, they are often short on practical detail, so your response is likely to be critical. If you want to hold their interest you will need to deliver your criticism constructively at a time when they are receptive – then they will find it useful. In return, this person will inspire you to broaden your horizons and have more fun.

With Mars in Capricorn

This is a strong and harmonious contact. The Mars in Capricorn person takes the lead in a practical, businesslike way. They have plenty of initiative, and when they feel confident they know how to take control and make things happen. You should find it easy to fit in with their well-organized approach, and your practical, thoughtful response will give them extra confidence. They can establish something strong and workable, while you attend to the details and make it work even better. This applies whether you are running a business, raising a family or making love.

With Mars in Aquarius

This person's approach is detached and intellectual, and they have a [illegible] might find their ideas inter-

With Mars in Pisces

This could be a case of opposites attract, or you may find yourself in conflict with this person. The Mars in Pisces person has strong, changeable feelings. When they feel confident they are relaxed and easy-going, but if they are not feeling so good they can be very scattered and confused. To attract this person you need to be sensitive to their moods. If you can do this, they will find your earthy, sensible advice helps them clarify their thoughts. In return, this person will help you to relax and get in touch with your feelings. If you are close, there is a strong sexual chemistry here, and you will find this person very seductive.

Famous Personalities with Venus in Virgo

Emily Brontë
Pauline Bonaparte
Lawrence of Arabia
Ginger Rogers
Lucille Ball
Robert Mitchum
Rita Hayworth
Peter Falk
Sophia Loren
Brigitte Bardot
Julie Andrews
Natalie Wood
John Lennon
Mick Jagger
Roger Moore
Cliff Richard
Wesley Snipes
Robert de Niro
Robert Redford
Julia Roberts
Heather Locklear
Rosanna Arquette

Ingrid Bergman
Catherine Deneuve
Sarah Ferguson
Melanie Griffith
Anjelica Houston
Martina Navratilova
Sylvia Plath
Linda Ronstadt
Diana Rigg
Luciano Pavarotti
Kate Winslet
Bob Geldof
Julio Iglesias
Ken Kesey
Jerry Lee Lewis
Fidel Castro
Carrie Fisher
Pierre Trudeau
Robin Williams
Antonio Banderas
Emmanuelle Beart

19
Venus in Libra

Charm is a way of getting the answer 'yes' without having asked any clear questions.
ALBERT CAMUS

You like your relationships to be pleasant and agreeable. Libra, the sign of the scales, is into harmony – polite, tactful, conscious of the needs of others, and able to compromise in a graceful and attractive way. This is how you prefer to conduct yourself, and you are attracted to others who share these qualities. For you, relating to others should be a refined and civilized activity without unnecessary crudeness or conflict.

Hugh Grant is a good example. It has been said that he relates to others with 'delicacy, refinement and sophistication'. Burt Lancaster was an actor of similar ilk, who was said to have 'a certain gentleness and sensibility…You have to give him a reason for everything…Once you do, he's easy to handle.'

You like to communicate in a clear, rational way with those you care about. For a well-balanced relationship both people must have a fair-minded attitude and this requires a certain detachment. However, sometimes you may become detached to the point where others find you a bit on the cold side – very nice and polite, but lacking in warmth. At times you might also find it difficult to reconcile being courteous with being truthful. Politeness can be used to cover up the truth, and if you do this some people may see you as being phoney. This is one of the contradictions of Libra – am I really that nice, or just politically astute? Or are they one and the same thing anyway?

Obviously you will take pleasure in socializing. Meeting new people and establishing harmonious relationships is an activity you can raise to an art form. You enjoy communicating with others, and your thoughtful, diplomatic style will help you develop your social skills to a highly refined level.

While the art of socializing is definitely your domain, you may also have a love of art generally. You like to be surrounded by beauty and harmony, and your taste in the arts will reflect this.

The sense of fair play you so value in your personal relationships will extend to a broader concern for social justice. Whether or not you actually pursue this depends on other factors in your chart, but you certainly possess a higher social awareness. Ideally you would like to live in a world which is peaceful, fair and civilized.

There are several celebrities with Venus in Libra who have followed this through. Using part of the proceeds from *Diamonds are Forever*, Sean Connery set up the Scottish International Education Trust in 1971 in order to help potential achievers from poor backgrounds. Similarly, Richard Gere established the Gere Foundation for Tibetan causes.

In love you tend to be on the conventional side. You like the contact to be pleasant and harmonious. While you may be quite romantic and sentimental, you won't necessarily enjoy a deep, full-on, emotional contact. Communication is all-important, and you like it to be gentle and thoughtful.

The Woman with Venus in Libra

You have a very gracious and charming personality, with a natural ability to create peace and harmony in relationships. Your manner is friendly and considerate towards all people, but you know how to maintain just the right distance to suit the occasion. You have a fine appreciation of art and culture, and in your dress and manner you present yourself with sophisticated elegance. You are a tactful person with an awareness of social etiquette, and you always treat people fairly. A good example of this was actor Grace Kelly, described by Bing

relationships you are a thoughtful,

courteous nature, and a good sense of diplomacy which allows you to relate easily to most people. Apart from the pleasure this brings, it is generally useful for getting on in the world. You are a man of taste who appreciates balance, harmony and beauty in people and art.

You idealize a woman who expresses these qualities in a highly refined way – someone who is cool and sophisticated, with a pleasant social demeanour. This is a cultivated person who carries herself with grace and elegance. She has a good and fair mind, with a sense of justice and a genuine concern for others. She loves beauty and harmony in all things, and may take an active interest in the arts. This is a beautiful and intelligent woman, a good friend and a cultured lover.

Your Compatibility with the Twelve Mars Signs

How will your Venus in Libra personality respond to the other person's approach? Before reading on, turn to the back of the book and look up his or her Mars sign. Then read the relevant section below.

With Mars in Aries

Aries and Libra are opposite signs, so there will be a strong attraction, but there is also potential for disagreement. The Mars in Aries person has a very direct, assertive approach which you may find tactless and offensive. On the other hand, you might feel they are refreshing and exciting to be with. At least this person is honest and sincere, and that is something you appreciate. If you really like them, you will help to smooth their rough edges, particularly in social situations. In return, they will inspire you to be more open and spontaneous, lustful even. This person could bring out the beast in you.

With Mars in Taurus

When this person really wants something, or someone, they set their course with great patience and determination. They are very down-to-earth, and, once moving, pretty well unstoppable. You may find them a bit heavy-handed at times, but you will also appreciate how richly productive their efforts can be. This person has financial prospects, and could offer you stability and security. They may also enrich your life with new sensual experiences. In return, you might help them develop a more refined taste and sophisticated social awareness.

With Mars in Gemini

The connection here is strong and harmonious. When this person is fired up, their mind becomes sharp and active. They have a passion for communicating and sharing ideas which you find very stimulating and, if you have common intellectual interests, you will never run out of things to talk about. This person is forever coming up with new ideas, and your thoughtful replies will help bring a balanced perspective into the discussion. If you are tired of intellectual conversations and want to move on to something a bit more raunchy, erotic correspondence with this person is a definite possibility.

With Mars in Cancer

This connection is quite challenging. The Mars in Cancer person expresses their passion with great strength and feeling. Their energy is very emotional and quite primitive, although they are capable of expressing it with tenderness and sensitivity. This person may lack refinement, but they could also captivate you with their powerful feelings – that is, if you haven't already beaten a polite but hasty retreat. When this contact works, it is very stimulating. The Mars in Cancer person can be very seductive, and could help you get in touch with your feelings. In return, you will bring intellectual refinement to the relationship.

[illegible]

They will make you laugh and feel [illegible] and refinement to the relationship. All in all you could make a very stylish couple, with a talent for entertaining.

With Mars in Virgo

This person's approach is practical and analytical. They become totally absorbed in whatever task they have set themselves, focusing their mind on every detail and trying to get it just right. While you may appreciate their quest for perfection, you might find they are too critical and worrisome, and a bit anti-social at times. If you really like them, you will encourage them to take a lighter, more balanced attitude in their work and social life. In return, they could inspire you to think more deeply and adopt a more pragmatic attitude.

With Mars in Libra

This is a strong, highly compatible connection. Essentially it is an intellectual contact which begins with respectful friendship. The Mars in Libra person will approach you on your own level – polite, thoughtful and fair-minded. They may be argumentative at times, but never unreasonable, and your friendly, detached responses should keep things on an even keel. This is a sensitive, civilized person who could quite possibly charm your socks off – and maybe more.

With Mars in Scorpio

The Mars in Scorpio person has a deep, primitive passion which could well be hidden under an ordinary-looking exterior. Don't be fooled by their pleasant manner. Not that this person is nasty, but they are very intense, and prone to jealousy and possessiveness once they become attached. Still, they are extremely loyal and also very seductive, and you just might find them fascinating enough to follow, like Beauty with the Beast. In return, you could help them develop a more sophisticated social awareness.

With Mars in Sagittarius

The connection here is quite compatible. When this person feels confident they express themselves with warmth and sincerity. They have a philosophical side to their nature and a good sense of humour. If you agree with their philosophy and find their principles sound, you will value their company and enjoy sharing ideas and conversation. They will inspire you and make you laugh – exploring life with this person will be a pleasure. When they get a bit carried away you will help them regain a balanced perspective. If you are close, you will find this person can be very lustful.

With Mars in Capricorn

Libra and Capricorn are two very different signs, so this contact is quite challenging. The Mars in Capricorn person wants to take control and do things their way. They are very practical and well-organized, and have probably thought things through carefully, so there is a good chance their way will work. But from your point of view, this is simply not the way you go about things in a relationship. Civilized people discuss things, share ideas, compromise when necessary and come to an agreement. You don't like being controlled and pushed around even if the other person *is* right. Perhaps other factors in your charts will modify this contact. If you *do* like this person you will find their self-assured strength to be helpful when you are indecisive. In return, you can help them refine their social skills.

With Mars in Aquarius

This is a strong, harmonious connection. When the Mars in Aquarius person is fired up they have a passion for communicating ideas. You

This person's approach is sensitive and easy-going. They can create an atmosphere which is quite enchanting, helping you to relax and stimulating your feelings and imagination. You will appreciate the peace and harmony which flows from this, and will enjoy adding your own graceful touch. However, when this person is not feeling good they will be surrounded by confusion. If you really like them, your affectionate, cool-headed response will help them regain their clarity of mind and a sense of peace.

Famous Personalities with Venus in Libra

Queen Elizabeth I
Mary Shelley
Oscar Wilde
D.H. Lawrence
F. Scott Fitzgerald
Burt Lancaster
Grace Kelly
Sean Connery
Woody Allen
Robert Vaughn
Prince Charles
Linda Evans
Michael Douglas
Patrick Swayze
Richard Gere
Boris Becker
Bill Clinton
Hugh Grant
Angie Dickinson
Britt Ekland
kd lang
Princess Margaret
Billie Jean King
Sean Young
Claudia Schiffer
Loretta Swit

Viviene Leigh
Charles Bronson
T.S. Eliot
Pablo Picasso
Charlton Heston
Dorothy Parker
Lulu
Bo Derek

20
Venus in Scorpio

To err is human; to forgive is not our policy.

ANON

In all your relationships you are deeply emotional, and you become strongly attached to those you care for. You have very sensitive feelings, and are naturally attracted to others with whom you can share this sensitivity in a deep, committed relationship. You are more likely to have a few close friends than a large number of superficial acquaintances, and you tend to form long-lasting relationships with the people who are important to you. You show great loyalty to these people, but you also have the ability to break from a relationship if you feel it has run its course.

Because your own feelings run so deep, you have an acute awareness of the deeper side to other people's natures. Like a sleuth, you seek to uncover their true motives. Jodie Foster expressed this when she said, 'The best thing that my education gave me was the opportunity to deepen my knowledge, and to understand how to read beneath the surface.'

If your judgement is good, you will be very canny and shrewd in your dealings with all kinds of people. If not, you may become paranoid, reading your own fears into their feelings and actions.

Because you are so sensitive, all your emotional experiences in relationships are intensified. When you love, you do so intensely, and when you commit yourself it is serious business. In love, you have the capacity for the most profound feelings of joy and emotional bliss. You will also feel the negative emotions very powerfully. As you become so deeply attached in relationships, your entire emotional being can be invested in the other person. If someone else arrives on the scene you may become extremely jealous, and if the one you love betrays you it will seem as if your whole world has collapsed.

While some people have no inclination for such emotional extremes, for you it is worth risking the pain to experience the bliss. In fact, at times you would choose a deeply painful relationship over a mediocre one. What you love is the emotional intensity. Deep down you also realize that by going through painful and difficult experiences in a relationship you grow stronger, and if you survive with your ability to love and trust intact you will have tremendous emotional strength and a deep capacity for a very fine quality of love.

After making *Dead Man Walking* with her partner Tim Robbins under difficult circumstances, Susan Sarandon said 'It was tough, but we are in a much better place now than we were before we started.'

Your deep, emotional sensitivity makes you vulnerable, and you feel the need to protect yourself. Consequently you are very private about your feelings, and may hide them beneath a bland exterior. This secrecy towards the world at large further intensifies the special bond of intimacy you have with the one you love and trust. Negatively, you may try to protect yourself by manipulating the other person to avoid being hurt. But in the process of avoiding pain you would also miss out on that deep, trusting connection you value so much.

In love you are romantic and passionate. For you it is virtually impossible to separate love from sex – particularly the deep emotional experience of sex. If the connection is clear and relaxed, your contact with the other person will be strongly psychic.

The Woman with Venus in Scorpio

Your beauty stems from your emotional power. Your feelings are so

relationship you value. You are very sensitive and perceptive. If you really like yourself, and your power doesn't go to your head, you could be the perfect combination of supportive partner and alluring lover.

The Man with Venus in Scorpio

Venus in Scorpio is a very feminine symbol – extremely sensitive and powerfully emotional. If the women with Venus in Scorpio are wary of expressing this openly, then the men are doubly so. This makes you a very private person when it comes to relationships. Whether in your career or personal life, you need to trust someone before you can work closely with them. Developing this trust takes time, but once it is established you make a very true and faithful partner.

You idealize a woman who is intensely passionate and emotional. There are no half measures with this person – she is captivating. She may be physically beautiful, or quite ordinary looking. It doesn't really matter because the intensity of your feelings enables you to see beauty in any woman with whom you feel a strong emotional connection. This is definitely *not* a faint-hearted person. She is strong, intelligent and seductive. There is a powerful sexual chemistry between you, and your commitment to each other is sacred.

Your Compatibility with the Twelve Mars Signs

How will your Venus in Scorpio personality respond to the other person's approach? Before reading on, turn to the back of the book and look up his or her Mars sign. Then read the relevant section below.

With Mars in Aries

When this person gets fired up they become extremely passionate. No doubt this would interest you. However, you could find them quite insensitive at times, and they may not be willing to give you the level of commitment you like. If you are still interested and can forgive their blunt manner, a relationship with this person could be very exciting, and they could introduce you to new experiences, both intellectual and sexual. Just remember not to tie them down too much.

With Mars in Taurus

The attraction here can be very strong, but as Taurus and Scorpio are opposite signs there is likely to be some conflict. While you value the earthy strength of this person, at times you may find them a bit pushy and judgemental. Neither of these signs is known for its flexibility, so if you do become locked in battle with this person it is a difficult cycle to break. However, if you basically get on well the attraction will be very strong. While they keep the relationship on a practical, steady course, you will infuse it with deep feeling and romance. If you are close, there will be a potent sexual chemistry between you.

With Mars in Gemini

When this person is switched on they have a great deal of restless nervous energy which they need to express. They may have a passion for playing games, driving cars, or creating art, but most likely their outlet is through communicating ideas. When they are confident and in full flight, this person can show a remarkable level of skill and dexterity. If you are interested in the things they talk about, they will keep you amused and entertained for hours on end. If not, you will find them boring and superficial, and their restless energy will drive you nuts.

When you *do* get on well with this person, they will inspire you to lighten up and be more detached. In return, you could seduce them [illegible] state of mind.

[illegible]

at times psychic. You will feel extremely caring and [illegible] each other, and your bond will be very private and full of feeling. Given the right mood, it is a question of who will seduce whom.

With Mars in Leo

This contact can be very stimulating, but as Leo and Scorpio are such different signs there is likely to be friction. When the Mars in Leo person feels confident they are very full-on and outgoing. You may find them overbearing and insensitive, in which case you will withdraw. On the other hand, if you find them attractive, they will inspire you to be more open and spontaneous – theatrical even. In return, you could help them develop a more subtle and sensitive approach. Just remember to stroke their ego when they need it. If you are close, this can be a very passionate and exciting connection.

With Mars in Virgo

The connection here is quite complementary. You will value this person's thoughtful, serious energy, and their down-to-earth approach can provide a healthy balance to your powerful moods and feelings. You will also appreciate their desire for thoroughness and attention to detail. This person could inspire you to be more organized and practical, while you help them become more sensitive to feelings. Together you could build a very private and caring relationship. If you are close, the connection will be very physical and emotional.

With Mars in Libra

This person has a strong desire to relate to others, and their approach is quite rational and detached. Usually they are very polite and charming, but sometimes they can be argumentative. While you may value their courtesy, you have no interest in polite diplomacy which hides a deeper truth, so if you really like this person you will probably prefer their company when they are argumentative – then it will be more intense. They will challenge you to take a detached, rational view, and also to be more sociable. In return, you will encourage them to look beneath the well-mannered surface of their social life, and perhaps get into something a bit more primitive.

With Mars in Scorpio

The connection here is very powerful. This person has strong emotional energy, and you will pick up on this whether they show it openly or not. The intensity of their feelings will attract you, and you will admire their thorough and purposeful approach. Once this person makes a commitment, they follow it through with great determination. If you decide to join forces, you will make a powerful couple. The bond between you will be private and intensely emotional. You will be very protective of each other, and extremely loyal. The sexual chemistry here is wickedly potent.

With Mars in Sagittarius

When this person feels confident, they are warm, generous and outgoing. There is a philosophical side to their nature, along with a sense of humour, and the adventure of life turns them on. You will admire their strength and honesty, although sometimes they will seem a bit blunt and tactless. If you really like this person, they will make you laugh, open your eyes to new horizons, and generally bring excitement and passion into your life. In return, you will encourage them to look beneath the surface and become more sensitive and shrewd in their dealings with others.

With Mars in Capricorn

The connection here is quite complementary. You value this person's strength and self-discipline, and their down-to-earth approach can

With Mars in Aquarius

This is a challenging contact. When this person feels confident, they are very much the passionate intellectual, and sometimes their ideas are quite unconventional. They may strike you as an eccentric intellectual with their head in the clouds. On the other hand, if you respect their mind you will find them interesting, even inspiring. If you really like this person, you will enjoy their honest, friendly nature. They will stimulate you to be more broadminded in your outlook, and perhaps also in your sexual habits. In return, you can help them come down out of their head, get more in touch with their feelings, and relax.

With Mars in Pisces

The connection here is very harmonious. When this person feels confident, they express themselves with a combination of strength and sensitivity which you should find most attractive. They have an easy-going, sympathetic nature, and when they are feeling vulnerable they will bring out your protective instincts. If you get stuck in a bad mood, they will help you relax and let go. When they feel lost and confused, you can help them regain their clarity and focus. If you are close to this person you will have a strong emotional connection and, given the right mood, an easy sexual rapport.

Famous Personalities with Venus in Scorpio

Martin Luther	Tracey Ullman
Marie Antoinette	Sigourney Weaver
Jane Austen	Stephanie Powers
Franz Liszt	Larry Flynt
Mahatma Gandhi	Bruce Lee
Kirk Douglas	Jim Morrison
Grace Slick	Annie Lennox
Goldie Hawn	Neil Young
John Cleese	Richard Dreyfuss
Jodie Foster	Patti Smith
Hillary Clinton	Steven Spielberg
Susan Sarandon	Demi Moore
Ava Gardner	Gwyneth Paltrow
Linda Evans	Leonardo di Caprio
Bette Midler	Sophie Marceau
Winona Ryder	Peter Cook

21

Sagittarius

The only true happiness comes from
squandering ourselves for a purpose.

JOHN MASON BROWN

In your relationships, you are playful and outgoing, often with a mischievous sense of humour, and you feel strongly attracted to people who share these characteristics. Your manner is naturally open and friendly, and you express your feelings with honesty, warmth and generosity. As Vidal Sassoon said of Kim Basinger, 'She's a positive, positive delight.'

For you, a pleasurable relationship is like a journey of discovery. This could be a physical journey, where you travel together discovering new people, places and cultures. You might also share a sense of fun and pleasure through recreation, particularly outdoor games and adventure sports.

On the other hand, your journey may be a purely philosophical one in which you explore the meaning of life. Perhaps you share with your friends a serious interest in spirituality, art or literature, or maybe you simply enjoy talking about people and what makes them tick. French actor Gerard Depardieu has a great love of literature which has inspired him in some of his best work. He also enjoys just sitting around gossiping, particularly about art and culture. Every new experience brings pleasure, and every new person opens up another horizon. You need this sort of challenge and stimulation to enjoy your life.

Because you like meeting new people you will make many friends, but you may find it difficult to commit yourself to one person. You love your freedom, and you look forward to new relationships with pleasure and optimism. You are idealistic about relationships, so it may be easier to keep moving on towards that ideal rather than settling down and dealing with the not-so-exciting everyday reality of a committed relationship. Even if you *do* settle down, there will be a side of your nature which longs to escape, if only for a while. You may do this through art, study or travel, but more than likely you will maintain an active social life so there are always new people to meet.

One of your most charming qualities is your sense of humour. When you feel good about yourself you are very warm and jovial. You really have the ability to cheer people up. You are able to get them laughing, talking and generally having a good time. In this mood your energy is infectious and uplifting. As Gerard Depardieu said, 'I love to hear people laughing.'

As a rule, you are a broad-minded person who is tolerant of other people's faults and weaknesses, but your principles are still very important to you. If a friend violates one of these important principles, your love and affection for them can disappear very quickly.

In love, you express yourself with warmth and passion, sometimes quite extravagantly. Honesty is very important to you, but you also have a very humorous side. Making love is a game of pleasure which you play with enthusiasm. It is also a journey of discovery to be shared with a friend.

The Woman with Venus in Sagittarius

Your beauty lies in your bright, generous personality. When you feel good, you express yourself with a warmth and optimism which inspires other people and makes them laugh. You have a playful sense of humour, but there is also a more serious side to your nature. You are a traveller and philosopher, and on your journey through life you have met many different people and thought seriously about your experiences. As a result, you have developed a strong set of principles and ideals, and underneath that outgoing character lies a person with

[illegible] feeling good, you strike

[illegible]

sign, so you should feel quite comfortable [illegible]

nature. You relate to others in a friendly, outgoing way. You like to be frank and direct with people, and although your bluntness may sometimes offend, your natural warmth and generosity will usually put matters right.

You have a philosophical bent, and a love for sport, travel and adventure, and all the places you have been and the people you have met provide ample material for the many stories you enjoy telling. Basically you are sincere and honest, but your tendency to get carried away with the enthusiasm of the moment may sometimes lead you to exaggerate just a little.

You idealize a woman who could be a friend and companion on your journey. This person is a mate, a pal – someone you can muck around with. She has a spirit of adventure and a sense of humour, and together you go places, explore and have fun. She is also someone you can talk to about life. She enjoys discussing ideas and beliefs, and will take pleasure in sharing your dreams and aspirations. This is a warm, passionate person with whom you can share a vision.

Your Compatibility with the Twelve Mars Signs

How will your Venus in Sagittarius personality respond to the other person's approach? Before reading on, turn to the back of the book and look up his or her Mars sign. Then read the relevant section below.

With Mars in Aries

The connection here is harmonious and high-spirited. When this person feels confident, they come straight to the point and say what they want. If you like them, you will respond with equal warmth and honesty, and the feelings between you will ignite quickly. You could go on an adventure with this person, and have plenty of fun along the way. The contact is playful and very passionate, and you should have no trouble in maintaining the excitement with humour and a bit of friendly competition. When it comes to sex, this contact is very high-spirited and lustful.

With Mars in Taurus

This person's approach is persistent and down-to-earth. They want to build something solid and productive in a relationship, and although they may be slow to get going, once they set their course they are very determined. If you are looking for some stability in your life, this may be just the right person, but you will need to compromise your love of freedom to some extent. This person could provide you with comfort and security, and in return you will encourage them to let go and have fun. Perhaps they could earn the money and you could spend it. If you are close, the contact is very passionate and physical.

With Mars in Gemini

Gemini and Sagittarius are opposite signs, so this contact can be very complementary, but there is likely to be some friction. When this person is turned on, they have a high level of nervous energy and a strong desire to communicate. If you are interested in what they have to say, you will find their company very stimulating. You can inspire each other with jokes, games and conversation, feeding each other's passion in a competitive but friendly way. When it works well, this contact is very lively and entertaining, and the friction between you could be extremely pleasurable.

With Mars in Cancer

This person has very powerful, sensitive feelings. Although their passion is strong, they need to feel secure before they will express it openly and confidently. If you really like them, make an effort to be sensitive to their moods – this will help them feel safe and relaxed. You

[illegible] more often than you would

[illegible]

centred at times, but if you like them [illegible]

along by their enthusiasm. This contact is passionate and playful, and as long as you can give the loyalty this person demands, you will find them very warm and generous. Make sure you stroke their ego on a regular basis. You may share an interest in sport, or perhaps the theatre. If you are close, this is a very lustful contact.

With Mars in Virgo

This connection can be very stimulating, but as Virgo and Sagittarius are quite different signs there is likely to be some conflict. While your preferred style is grand and passionate, the Mars in Virgo person takes a careful, practical approach, paying great attention to detail. While you may find them infuriatingly fussy and critical, they might think you are careless and unrealistic. On the other hand, they might inspire you to become more discriminating and generally clean up your act. In return, you could help them open up and have some fun.

With Mars in Libra

This person has a strong desire to relate to others, and as a rule they are very courteous and fair-minded. Sometimes they will be argumentative, but even then they will remain open to reason. You should find their conversation to be pleasantly stimulating, and if you are really interested in what they have to say you will respond with warmth and pleasure. If you start going over the top, this person can help you calm down in a friendly and tactful way. In return, you could inspire them to laugh and enjoy themselves. The connection here is friendly, pleasurable and quite complementary.

With Mars in Scorpio

This person's passion runs deep. They are very sensitive and emotional, but they won't always express it openly. You have to tune in to their feelings, and learn to read the signs, but even when you are right don't expect them to admit it – not until they trust you, anyway. If you really like this person, you will find them intense and captivating, but you may have to compromise your need for freedom. The Mars in Scorpio person wants total commitment, and may become quite jealous and possessive. Sexually they can be very potent and seductive.

With Mars in Sagittarius

The contact here is strong and harmonious. When this person is fired up, their approach is warm, friendly and expansive. They have a strong sense of adventure, with a desire to explore new places and ideas, and if you are on the same wavelength you will love to go with them. This is someone you can travel the world with, sharing the same dream. Together you could have a lot of fun, while at the same time gaining a deeper insight into yourself and life generally. This can be a warm, friendly contact, full of passion, laughter and new experience.

With Mars in Capricorn

When this person wants something, or someone, they go about it in a very businesslike way. You may find them a bit controlling at times, but you will admire their enterprising approach and their ability to get things done. While you can inspire them with ideas and insight, they could teach you the wisdom they have gained through hard experience. If you want to build a strong partnership, this could be the person to do it with. However, to make the best of the relationship you would need to discipline your energy and settle in for the long haul. On a personal level, this connection has an earthy passion.

With Mars in Aquarius

This person's approach is friendly, but quite detached. They have a passion for ideas, and a strong desire to communicate. If you are inter-

With Mars in Pisces

This contact can be very stimulating, but also challenging. The Mars in Pisces person is very emotional and sensitive, but not always as direct and outgoing as you would prefer. If you really like this person, you will learn to read their moods and know how to respond. Sometimes your bright, cheerful energy is just what they need, but at other times, if you can't relate sympathetically, you would do better to leave them alone. When this person feels confident, their combination of strength and sensitivity can be most seductive. They are good at creating a romantic atmosphere, and you may find yourself quite carried away.

Famous Personalities with Venus in Sagittarius

Jonathan Swift
Emily Dickinson
Lewis Carroll
Mark Twain
Robert Louis Stevenson
Rudyard Kipling
Winston Churchill
Albert Schweitzer
W.C. Fields
Al Capone
Cary Grant
Howard Hughes
Johnny Carson
Paul Simon
Jimi Hendrix
Pamela Stephenson
Frank Zappa
Gerard Depardieu
Michael Landon
Kim Basinger
Kevin Costner
Whoopi Goldberg
Joan Baez
Farrah Fawcett

Sally Field
Jane Fonda
Germaine Greer
Eartha Kitt
Linda Lovelace
Mary Queen of Scots
Roseanne
Tina Turner

Alan Alda
David Bowie
Evel Knievel
Bela Lugosi
Quentin Crisp
Kahlil Gibran
Jamie Lee Curtis
John Malkovich

22

… in Capricorn

No one who is in a hurry is quite civilized.

WILL DURANT

In love and friendship you are a practical, hard-working person with a strong sense of commitment to those you care for. You are attracted to partnerships you think will be productive and rewarding, and you take pleasure in helping friends and loved ones achieve their practical goals in life. In a relationship, you like to build something concrete, otherwise you don't see it as a valuable experience. This is serious business, and you are looking for real results.

As well as being practical and supportive in relationships, you have an enterprising touch and a flair for business which could flourish in partnership. Naturally you are attracted to someone who also possesses these qualities – a strong, well-organized person who is not afraid of hard work. Together you can build something lasting and worthwhile, whether this is a family or business. In the process of working towards these goals, you will achieve a position of standing within society, along with the wisdom of experience.

Wisdom and maturity are qualities you value, both in yourself and in others. Consequently, you may be attracted to relationships with older people. This was the case with Clark Gable, who once remarked, 'The older woman has seen more, heard more, and knows more than the demure young girl…I'll take the older woman every time.' James Dean also enjoyed numerous affairs with older women.

Singer Maria Callas had two significant relationships, both with older men who were also wealthy and powerful. Looking back on the second of these – with shipping magnate Aristotle Onassis – she commented, 'We were doomed, but oh how rich we were...'

Your practical approach to relationships might tempt you to marry solely for reputation or material security. While you won't necessarily follow this through, no doubt it has crossed your mind at some stage. Certain people may believe that romance is somehow enhanced by poverty, but you know it is much more likely to blossom with the help of a healthy bank account. However, if it's only your partner's bank account which is healthy, there will be something missing for you. Your greatest pleasures will always come through a sense of your own achievement, through time, experience and hard work.

As a general rule with Capricorn, if you put in the time and effort, the older you get the better life becomes. With Venus in Capricorn, this can mean that your beauty develops as you grow older. For example, many people think that Paul Newman has become more attractive with age. Also, because you value maturity, you may develop a particular love for objects of age and beauty – like castles, antiques or fine wines.

You are very physical when expressing affection, but you need to feel secure with someone before showing your feelings. Initially, you will hide behind a well-mannered reserve while you check the other person out. Your preferred style is carefully measured, and you like to feel in control every step of the way. When you *do* allow yourself to love someone, your commitment is strong and you can express your feelings with great power and sensuality.

The Woman with Venus in Capricorn

Your beauty lies in your strength and wisdom, and this naturally increases as you grow older. In public, you conduct yourself with a dignified elegance which commands respect from others. Your attractiveness is enhanced by formal occasions which highlight your well-mannered beauty. You really are a very classy person. In private, you are cautious before letting go of your feelings. This natural reserve is part of your attractiveness – you take love seriously and don't give your-

... something to be valued. Your

you, a ...

hard work. You have a natural sense of decorum ...
in good stead in formal and business relationships. On the personal level, you need to feel secure with someone before you open up and express your feelings.

You idealize a woman who has real strength of character – a practical person with poise and self-control. Noel Coward described his ideal woman as 'coolly elegant, seemingly well-bred, her manners impeccable and her conduct, at any rate in public, above reproach'. Whether you introduced her to a king or a slave, she would handle it with well-mannered self-assurance. This is someone you can depend on to work hard at helping you build a strong relationship. At first she may be shy about expressing her feelings, but you would feel comfortable with this. You also like to take your time, and you find her natural reserve to be most attractive. If she did decide to give total commitment, she would be a dependable partner and a sensual lover.

Your Compatibility with the Twelve Mars Signs

How will your Venus in Capricorn personality respond to the other person's approach? Before reading on, turn to the back of the book and look up his or her Mars sign. Then read the relevant section below.

With Mars in Aries

The attraction here can be strong, but Aries and Capricorn are quite different signs so there is likely to be some friction. You may find this person's energy annoyingly tactless, even offensive. They have a strong individual streak, and once fired up they can be quite undiplomatic. If you really like this person, you will forgive their impropriety, or at least tolerate it. In fact, you may find they bring a new sense of excitement and passion into your life. In return, you will encourage them to direct their energy in a more sen-sible and dignified manner.

With Mars in Taurus

The connection here is very harmonious. When this person takes the lead, their energy is strong and earthy. You appreciate their desire for stability and material security, and your practical energies should blend well as you work steadily to build a relationship. Given time and effort, this could be a highly productive partnership. Just try not to turn into a pair of workaholics. On an intimate level there is an easy connection – very physical and sensual.

With Mars in Gemini

This person's approach is detached and intellectual. When they get fired up, they have a nervous, restless energy, and a strong desire to communicate. If you are interested in what they have to say, and believe they are capable of putting their ideas into practice, you will find this person to be interesting and stimulating company. In this case they will entertain, amuse, and give you ideas. In return, you could help them direct their energy in a constructive, down-to-earth way.

With Mars in Cancer

The attraction here can be very strong, but as Cancer and Capricorn are opposite signs there is likely to be some conflict. This person has sensitive feelings which they can express with great power. On the one hand you may be fascinated by their passionate moods, but on the other they will probably threaten that carefully-controlled, well-ordered world you like to share in a relationship. This person challenges you to take an emotional risk. If you are willing to be seduced by them, it will be a new emotional experience. But there needs to be give and take here, so this person must respect your limits, whatever they are. If you are close, and feel secure together, this could be a very potent sexual contact.

can encourage them to direct their energy in a [illegible] cal way, and you will also work hard to help them achieve their goals. On the personal level, the contact is physical and passionate.

With Mars in Virgo

The connection here is strong and harmonious. This person directs their energy in a thoughtful, practical way. You value their down-to-earth approach, and they should find your feedback to be helpful and supportive. Together you could build a strong, well-ordered relationship, whether this is business or personal. At the end of the day, when the work is done and everything is in its proper place, you could enjoy a contact which is strongly sensual and physical. Unless, of course, you decide to keep on working into the night.

With Mars in Libra

This is a challenging connection. While you like someone who knows how to take charge, the Mars in Libra person wants to talk things over and reach an agreement. They might be very gentle in their approach, or quite argumentative. When they are gentle you may see them as being weak, but this isn't necessarily the case – often it is just their manner. If you really like this person, they may show you what can be achieved through compromise and diplomacy. In return, you will help them take control when the time is right.

With Mars in Scorpio

The contact here is quite complementary. This person has strong, emotional energy, and when they really want something, or someone, they are very determined and persistent. You will value the thoroughness of their approach, and if their emotions begin to cloud their judgement, your cool-headed, practical response will be just what they need. Together you could create a strong, dependable partnership which will weather any storm, but this will take time. If you develop a close, trusting bond, your connection will be deeply emotional and strongly physical.

With Mars in Sagittarius

This person has a passion for travel and adventure, along with a philosophical bent. They are inspired by new ideas and new directions, and when things are going well they are warm, enthusiastic and fun to be with. If you like where this person is going, you will find them stimulating company, and they will inspire you to broaden your horizons. In return, your down-to-earth response will help them keep their feet on the ground and achieve their goals. When it comes to sex, this person can be very passionate, and may draw you out of yourself in a new and exciting way.

With Mars in Capricorn

The connection here is very powerful. When this person feels confident, they take control in a very businesslike way. They have a strong sense of their own authority, and it seems only a matter of time before they will achieve their carefully thought out ambitions. If you are on the same wavelength, you will take great pleasure in helping them, which will give them even more confidence in themselves. This is a good combination in business as well as on the personal level. On an intimate level the contact should be easy – very physical and sensual.

With Mars in Aquarius

This person becomes strongly attached to ideas, and may express them with great passion. They are very honest, and can also be quite eccentric. You will probably value their seriousness and sincerity, but whether you like them or not will depend on what you think of their ideas. If you value their mind, you will encourage them to take a sensible approach so they can put their ideas into practice. In return, they will inspire you [illegible] speak your mind without fear. If you like

[illegible]

can help them regain a clear, practical [illegible] help you calm down if you are feeling uptight. This person can be very seductive. If you allow yourself to relax and let go with them, you could be quite carried away.

Famous Personalities with Venus in Capricorn

Beethoven
Dostoyevsky
Harpo Marx
Noel Coward
Clark Gable
Lorne Greene
Paul Newman
Rock Hudson
Robert Kennedy
Elvis Presley
Peter Fonda
David Carradine
Danny de Vito
Greg Norman
Diane Keaton
Mikhail Gorbachev
Steven Spielberg
Michael Jordon
Monica Seles
Edith Piaf
Eva Braun
Princess Caroline
Faye Dunaway
Indira Gandhi
Lisa Marie Presley
Camille Claudel
James Dean
Henry Miller
Burt Reynolds
Neil Diamond
Placido Domingo
Brad Pitt
Prince Andrew

23
Venus in Aquarius

A friend is somebody who knows all about
you but likes you just the same.
CONTACT

In all your relationships you are honest, friendly and quite detached, and you are strongly attracted to people who share these qualities. You value the truth, and you like people who know and speak their mind. Even if you disagree with them, you would prefer they were honest with you than play deceptive games or pander to your ego.

You are drawn to people who relate to others with a detached, intellectual air. In relationships you are fairly independent, and you need the freedom to have as many friends as you like of either sex. You don't like intense emotions, particularly the clinging, claustrophobic kind – like jealousy, which overwhelms reason and makes it difficult for you to remain detached. Sometimes others may see you as distant and impersonal, but whether this is their problem or yours is a matter of opinion.

You like to feel you are a free-thinking person with your own individual style, and will enjoy sharing this side of your nature in relationships. Your friendly manner gives you the ability to get on with all types of people, no matter who they are or where they come from. In fact, some of your relationships may be quite unusual, with people who are decidedly different. You don't like to feel tied by convention – better to be unconventional and honest about it than to live a traditional lie. But then again, you might enjoy being traditional. Either way, it's your decision, and you take pleasure in choosing your own friends accord-

... whatever those values might be.

were freedom from ...

other.

You are a highly sociable person who enjoys communicating with like-minded people. This means you can work well with a group of people who have similar values. Whether it's an informal social gathering or a team of people working for a cause, you are able to adjust your goals to fit those of the group and promote friendly co-operation.

When it comes to love, it must be based on friendship. You are serious about your commitment and are not inclined to play games, but although you may be faithful you are also capable of changing your mind. Communication is all-important, and you like it to be open and honest.

The Woman with Venus in Aquarius

Your beauty lies in your friendly, outgoing manner and your strong individual spirit. You relate to others with a detached honesty which allows them the space they need to be themselves, and gives you the freedom to be *your*self. You may be conservative or radical, avant garde or classical. Your dress, your manners, your taste in art – all these things you choose yourself. You may even decide on an eccentric combination – like Cybill Shepherd, who has been described as a '90s feminist crossed with an old-fashioned Southern belle'.

One of your most attractive qualities is your fair-mindedness which gives you the ability to rise above petty games and jealousy. You can relate on equal terms with anyone you choose, putting them at their ease with friendly conversation and listening with genuine interest to what they have to say. You are a person with a concern for your fellow human beings which transcends narrow, personal self-interest.

The Man with Venus in Aquarius

In relating to others, you are open-minded and independent. You enjoy socializing with a wide variety of people, some of whom may be quite eccentric. Communicating and sharing ideas brings you pleasure, but while you may take the conversation quite seriously, you like to maintain a cool, objective view of things.

You idealize a woman who is intelligent and unusual. She is a free-thinking person who goes her own way, lives her own life, and has many different, interesting friends. Although you have complete faith in her honesty and commitment, you never quite know what is going through her active, original mind. She's a great person to talk to – serious, funny and often quirky. Even though she is quite cool and detached, she has a genuinely kind heart and a real concern for the welfare of others. This woman is different – a good friend and a faithful lover, but also a free-spirited individual with a mind of her own.

Your Compatibility with the Twelve Mars Signs

How will your Venus in Aquarius personality respond to the other person's approach? Before reading on, turn to the back of the book and look up his or her Mars sign. Then read the relevant section below.

With Mars in Aries

The contact here is quite complementary. When this person feels confident, their approach is open and direct. You appreciate their strong individuality, and won't feel threatened by their independent nature. Their level of passion may throw you a bit at first, but once you realize they are not trying to own you, you should find it refreshing and exciting. In return, you could help this person keep a cool head when they [illegible] upset. This could be a liberating contact between two

[illegible]

there is nothing you like less than [illegible] and own you. As Taurus and Aquarius are both fixed signs it could be quite a battle working out a compromise between your love of freedom and their desire for security. However, if you succeed this contact can be very stimulating.

With Mars in Gemini

The connection here is easy and harmonious. This person has a passion for communicating, and if your minds are on a similar wavelength you will find their company highly stimulating. They will amuse you with witty stories and entertain you with interesting conversation, and they are unlikely to feel threatened when you want to go and socialize with someone else. This could be a healthy friendship between two like-minded people who never run out of interesting things to talk about. On an intimate level, phone sex is a potential pleasure worth exploring.

With Mars in Cancer

This person has strong feelings which they can express with great power and sensitivity. While you may find them to be fascinating and different, you are unlikely to enjoy the depth of emotional attachment they desire. This person wants emotional security, but you like your freedom. They want a close, emotional bond, but you prefer to keep a certain distance. There may be other factors in your charts which help this to work, but you will need to compromise to some extent. If you get on well, this person will help you express your deeper feelings, while you encourage them to be more cool and detached in their relationships.

With Mars in Leo

Leo and Aquarius are opposite signs, so there is bound to be some tension in this contact – but there is also a powerful attraction. When this person feels good, they express themselves with a warmth and sincerity which you should find appealing. Their energy is strong and true, and although they may play games it is unlikely they are trying to deceive you – basically they want to have fun. If you like this person, you will find they draw you out of yourself and make you laugh. Just remember they are very proud, and won't appreciate hearing the truth if it hurts their ego – especially in public. When this contact works, there is a strong sexual attraction.

With Mars in Virgo

You will find this person's approach to be very thoughtful and particular. They are very earthy, and have a passion for analysing things. If your minds are on the same wavelength you will find their serious, intellectual bent to be quite appealing, although at times their private, analytical world may seem a bit claustrophobic. If you get on well, this person will inspire you to be more practical and down-to-earth. In return, you will encourage them to be more outgoing and sociable.

With Mars in Libra

The connection here is easy and harmonious. The Mars in Libra person has a strong desire to relate to people, and they are turned on by stimulating conversation. Their approach may be gentle and polite, but they can also be quite argumentative. Either way, your cool, detached response should be well received, and if you share similar interests the conversation will be enjoyable and stimulating. As a couple, you would work well together on social occasions. The Mars in Libra person will initiate conversation, whether it's polite chat or serious debate, and you will help to keep it flowing. This could be a strong and lasting friendship. If you are looking for something more intimate, ... the conversation in that direction.

those nasty, ...

possessiveness. There may be other factors in ...

you resolve this difference, but you will both need to be flexible. If you *do* get on well, the contact is highly stimulating. This person can be very seductive and they may inspire you to express your deeper feelings. In return, you could help them become more open and communicative in their relationships.

With Mars in Sagittarius

The connection here is easy and complementary. When this person feels confident, their energy is warm and enthusiastic. They have a passion for ideas and a strong spirit of freedom. As long as you have similar principles you will enjoy sharing new insights and stimulating conversation. If you get on well, this person could bring a new level of excitement and pleasure into your life. Lust and passion may be the order of the day. And if they get overheated, your cool head will calm them down pleasantly. This could be a most enjoyable relationship between two original people.

With Mars in Capricorn

This person has a well-organized, down-to-earth approach. You should admire their strong initiative, and if you think they are heading in a worthwhile direction you will offer friendly support. If they try to control you too much you won't enjoy it, and then you will need to make your position clear – that your freedom is important to you. If you get on well, this person could inspire you with the wisdom of their experience and their practical achievements. In return, you will help them with new ideas and encourage them to be more detached and sociable.

With Mars in Aquarius

This is a very powerful connection. The Mars in Aquarius person will approach you on your own level, and you should find it easy to respond to their open, honest passion. They have a strong, original mind and an independent spirit, and if you share a common interest the contact will be pleasurable and highly stimulating. While some people may think this person is a bit odd, you will probably find them quite interesting. Your conversation may simply be a friendly exchange, or it might show a serious concern for broader life issues. This could be a very strong friendship between two original, free-thinking people. If you are close, sex will probably be one of your favourite topics of conversation.

With Mars in Pisces

This person has a sensitive, easy-going approach. When they feel confident, they can create an atmosphere which is full of feeling and pleasurably seductive. You may find it easy to drift away with them, but you are unlikely to become as emotionally involved as they would want. While you may share a sense of idealism, you like to remain clear-headed about it. If you are close, this person will inspire you to really get in touch with your feelings. In return, you will encourage them to keep a clear head when it matters.

Famous Personalities with Venus in Aquarius

Nostradamus
Isaac Newton
Mozart
Robert Burns
Lord Byron
Frederic Chopin
Randolph Scott
Marlene Dietrich
David Niven
Frank Sinatra
Robert Wagner
Kim Novak

Yoko Ono
Alan Arkin
Karen [illegible]
Glenn Close
Morgan Fairchild
Margaret Mead
Carmen Miranda

Patricia Neal
Carole King
Gertrude Stein
Peter Gab[illegible]
Erica Jong
Christine Keeler
Billy Crystal

24
Venus in Pisces

I can't figure out where I leave off and someone else begins.

GEORGE McCABEE

In love and friendship, you tune in to other people through your super-sensitive feelings. You pick up easily on their vibes, and generally respond with sympathy. In relationships, your world is a sea of ever-changing feelings, and like a chameleon you change with it. Your fluid nature is extremely responsive, and even if you don't know someone very well you are still able to relate to them with sympathetic understanding. This makes you a very seductive person, and also seduceable.

Because of your changeable moods, some people may think you are fickle, or evasive. Perhaps you are. But it may just be that you find it difficult to put your feelings into words. The writer Anais Nin described her feelings for her lover June as follows – 'I can't think when I am with you. You are like me, wishing for a perfect moment, but nothing too long imagined can be perfect in a worldly way. Neither one of us can say just the right thing. We are overwhelmed. Let us be overwhelmed.'

You have an idealistic side to your nature which colours all your relationships. You enjoy fantasy and mystery, and are able to identify closely with those characters in books and movies who capture your imagination. Depending on other factors in your chart, you may actively create your own idealistic dream world and express this through the arts. Charles Dickens did this in his writing, often evoking sympathy through the image of the lost child, the waif. Fellow writer

Aldous Huxley said of him 'Whenever he is in the melting mood, Dickens ceases to be able to and probably even ceases to wish to see reality.'

have a bit of earth in your chart to bring

When you are feeling relaxed, you have the ability to help others relax too. In this mood you can be very healing in relationships. You are naturally attracted to other sensitive people with whom you can share a sympathetic and supportive connection. Although your feelings may be changeable, in love you are generally very devoted and caring.

The Woman with Venus in Pisces

Your beauty lies in your soft, sensitive, caring nature. You are a considerate person, easily touched by other people's feelings, and you respond readily to those in need. When you feel good about yourself, you have a dreamy, seductive aura. In this mood you are quite enchanting, and other people will be drawn to you. Sally Field once said that Joanne Woodward had a 'love and warmth that you instantly fall into'. Another actor with Venus in Pisces is Nastassja Kinski. It has been said of her that she 'combines the innocent look of an angel with the guilty appeal of a sex kitten'.

The Man with Venus in Pisces

Venus in Pisces is a very soft, feminine symbol. It gives you an easygoing sensitivity which you like to express in your relationships. You are a charitable person, always willing to see the better side of others. Sometimes they will disappoint you, but perhaps it is better to give them the benefit of the doubt, at least to begin with. Alternatively, you may hide your sensitive feelings to avoid being hurt.

You idealize a woman who is soft and impressionable, and extremely sensitive to the feelings of others. She has a particular sympathy for people who are lost or confused, and is able to put aside her own needs in order to help them. When you are uptight, she will be understanding and attentive.

She has a rich, creative imagination, which may be expressed artistically. Although her feelings are changeable, she is devoted and caring. This is a sensitive woman with an air of mystery which you find most attractive.

Your Compatibility with the Twelve Mars Signs

How will your Venus in Pisces personality respond to the other person's approach? Before reading on, turn to the back of the book and look up his or her Mars sign. Then read the relevant section below.

With Mars in Aries

This person's approach is open and direct. They are strongly independent, and their innocent, unthinking manner could make them appear insensitive at times. If you can forgive their abruptness, understanding they don't mean to be hurtful, you may find their energy quite refreshing. This person could bring excitement and adventure into your life – lust and passion perhaps. In return you might help them relax, and give support and understanding when they need it.

With Mars in Taurus

The connection here is quite complementary. This person has a very practical side to their nature which can nicely balance your romantic idealism. They are willing to work hard in order to achieve something solid in their life and relationships. At times they may be stubborn and judgemental, but your sensitive touch will help them soften their approach. This person offers stability and material security. With their

... you could have a

Gemini pe...

nicating. Their approach is quite detached, and they ...

stand things in a rational, logical way. Obviously they'll have trouble working you out. If you don't get on, they will find you puzzling and confusing, while you see them as being too intellectual and out of touch with their feelings. On the other hand, if you like each other there will be a mutual fascination for someone who is different and interesting. They could inspire you to put your feelings into words, while you help them slow down and relax.

With Mars in Cancer

The connection here is very harmonious. When this person feels confident they express their feelings with great power and sensitivity. If you like them, you will respond immediately, and your feelings will be in sympathy. You can achieve a very close emotional intimacy with this person – a rapport which is deeply caring, and at times psychic. This could be a very healing contact between two extremely sensitive people. On an intimate level, it is a mutually seductive combination.

With Mars in Leo

When this person feels confident, their approach is strong and forceful. At times you may find them self-centred and overbearing, but they can also be very warm and positive, especially when things are going their way. If you really like them, you will learn to deal with their more insensitive moods, and take pleasure in their warm-hearted enthusiasm when times are good. Don't hold back on the compliments if they are doing well. This person could give your life a lift. In return you will offer sympathetic understanding when they need it. When it comes to sex, this person can be quite theatrical, and very lustful.

With Mars in Virgo

This is a very stimulating connection, but as Virgo and Pisces are opposite signs there is bound to be some tension. This person's approach is thorough and down-to-earth. They can be very critical and fussy, but at times their sound, practical judgement will be just the thing you need to clarify your thoughts and help you make the right decision. On the other hand, if they are becoming worried and uptight, your gentle, easy-going mood will calm them down. For this contact to work well, you both need to appreciate the value of your opposing natures. If you do, your relationship could be very stimulating, both intellectually and sexually.

With Mars in Libra

This person is quite detached and intellectual in their approach. They may be charming and courteous, or quite argumentative. Naturally you will prefer their softer side, but even then you may sometimes find them a bit cool and distant. If you really like this person, their detached, objective attitude will help you maintain a clear-headed view of things. In return, you will encourage them to be more sensitive to feelings. Despite your differences, you may share an interest in the arts, or perhaps a genuine concern for a better society. If you are looking for intimacy, start with stimulating conversation.

With Mars in Scorpio

The connection here is very harmonious. This person has deep and powerful feelings. Although they won't always express them openly, you will pick up on them straightaway. If you like this person, your emotional response will be strong, and it should be easy for you to establish a close, trusting bond. Their intense feelings will provide a focus for your relationship, but if they get too wound up you will help them let go and relax. If you are close, there will be an easy sexual rapport and your bond will be strongly psychic.

can inspire you with their positive energy. [illegible] smooth their rough edges and support them when they need it. Sexually, this person can be very playful and lustful.

With Mars in Capricorn

The connection here is quite complementary. This person has a very practical, well-organized approach which offers a natural balance to your sympathetic idealism. If you are feeling a bit lost and out of sorts, they can be depended on to take charge and sort things out. On the other hand, when they are feeling overworked and over-wrought, you might seduce them into a more relaxed state of mind – or maybe state of undress. This person could probably do with less work and more pleasure. Use that imagination of yours.

With Mars in Aquarius

This is a strong-minded person with a passion for communicating ideas. Their approach is quite detached and their ideas can be very original. At times they may seem fairly distant, but if you are interested in what they have to say you will find their company stimulating. When they get stuck in their head, you can help them relax and get back in touch with their feelings. If you are close, you could share a sense of idealism which goes beyond the purely personal. However, if you prefer to keep it on the personal level, you may find this person has some very imaginative ideas when it comes to sex. Encourage them to talk about it.

With Mars in Pisces

The connection here is very strong. When this person feels confident they can express their feelings with strength and sensitivity, and you will find the mood they create is very seductive. If you really like them, it will be a pleasure to be seduced. As a rule, you will tune into each other's moods very easily, and together you could spend many enjoyable hours avoiding reality – watching movies, having sex, or discussing how you are going to save the world. This is a very easy-going and dreamy contact between two sensitive and imaginative people.

Famous Personalities with Venus in Pisces

J.S. Bach
Casanova
George Washington
Hans Christian Andersen
Charles Dickens
Vincent Van Gogh
Nikolai Lenin
Billie Holiday
Ella Fitzgerald
Hugh Hefner
Sidney Poitier
Joanne Woodward
Shirley MacLaine
Vanessa Redgrave
Jack Nicklaus
Barbra Streisand
Michael Caine
Richard Nixon
John Travolta
Nastassja Kinski
David Letterman
Tom Selleck
Colette
Michelle Pfeiffer
Drew Barrymore
Geena Davis
Laura Dern
Shannen Doherty
Queen Elizabeth II
Bridget Fonda
Patty Hearst
Barbara Hershey
Anais Nin
Camille Paglia
Diana Ross
Alice Cooper
Jack Kerouac
Edgar Allen Poe
Lee Majors
Rod Stewart
Mimi Rogers

Mars and Venus Sign Positions AD 1900–2000

1900	Mars in
1 JAN–21 JAN	CAPRICORN
21 JAN–28 FEB	AQUARIUS
28 FEB–8 APR	PISCES
8 APR–17 MAY	ARIES
17 MAY–27 JUN	TAURUS
27 JUN–10 AUG	GEMINI
10 AUG–26 SEP	CANCER
26 SEP–23 NOV	LEO
23 NOV–31 DEC	VIRGO

1900	Venus in
1 JAN–20 JAN	AQUARIUS
20 JAN–13 FEB	PISCES
13 FEB–10 MAR	ARIES
10 MAR–6 APR	TAURUS
6 APR–5 MAY	GEMINI
5 MAY–8 SEP	CANCER
8 SEP–8 OCT	LEO
8 OCT–3 NOV	VIRGO
3 NOV–28 NOV	LIBRA
28 NOV–23 DEC	SCORPIO
23 DEC–31 DEC	SAGITTARIUS

1901	Mars in
1 JAN–1 MAR	VIRGO
1 MAR–11 MAY	LEO
11 MAY–13 JUL	VIRGO
13 JUL–31 AUG	LIBRA
31 AUG–14 OCT	SCORPIO
14 OCT–24 NOV	SAGITTARIUS
24 NOV–31 DEC	CAPRICORN

1901	Venus in
1 JAN–16 JAN	SAGITTARIUS
16 JAN–9 FEB	CAPRICORN
9 FEB–5 MAR	AQUARIUS
5 MAR–29 MAR	PISCES
29 MAR–22 APR	ARIES
22 APR–17 MAY	TAURUS
17 MAY–10 JUN	GEMINI
10 JUN–5 JUL	CANCER
5 JUL–29 JUL	LEO
29 JUL–23 AUG	VIRGO
23 AUG–17 SEP	LIBRA
17 SEP–12 OCT	SCORPIO
12 OCT–7 NOV	SAGITTARIUS
7 NOV–5 DEC	CAPRICORN
5 DEC–31 DEC	AQUARIUS

1902	Mars in	1902	Venus in
1 JAN	CAPRICORN	1 JAN–11 JAN	AQUARIUS
[illegible]	[illegible]	11 JAN–6 FEB	PISCES
20 DEC–31 DEC	LIBRA	[illegible]	[illegible]
		7 OCT–31 OCT	LIBRA
		31 OCT–24 NOV	SCORPIO
		24 NOV–18 DEC	SAGITTARIUS
		18 DEC–31 DEC	CAPRICORN

1903	Mars in	1903	Venus in
1 JAN–19 APR	LIBRA	1 JAN–11 JAN	CAPRICORN
19 APR–30 MAY	VIRGO	11 JAN–4 FEB	AQUARIUS
30 MAY–6 AUG	LIBRA	4 FEB–28 FEB	PISCES
6 AUG–22 SEP	SCORPIO	28 FEB–24 MAR	ARIES
22 SEP–3 NOV	SAGITTARIUS	24 MAR–18 APR	TAURUS
3 NOV–12 DEC	CAPRICORN	18 APR–13 MAY	GEMINI
12 DEC–31 DEC	AQUARIUS	13 MAY–9 JUN	CANCER
		9 JUN–7 JUL	LEO
		7 JUL–17 AUG	VIRGO
		17 AUG–6 SEP	LIBRA
		6 SEP–8 NOV	VIRGO
		8 NOV–9 DEC	LIBRA
		9 DEC–31 DEC	SCORPIO

1904	Mars in
1 JAN–19 JAN	AQUARIUS
19 JAN–27 FEB	PISCES
27 FEB–6 APR	ARIES
6 APR–18 MAY	TAURUS
18 MAY–30 JUN	GEMINI
30 JUN–15 AUG	CANCER
15 AUG–1 OCT	LEO
1 OCT–20 NOV	VIRGO
20 NOV–31 DEC	LIBRA

1904	Venus in
1 JAN–5 JAN	SCORPIO
5 JAN–30 JAN	SAGITTARIUS
30 JAN–24 FEB	CAPRICORN
24 FEB–19 MAR	AQUARIUS
19 MAR–13 APR	PISCES
13 APR–7 MAY	ARIES
7 MAY–1 JUN	TAURUS
1 JUN–25 JUN	GEMINI
25 JUN–19 JUL	CANCER
19 JUL–13 AUG	LEO
13 AUG–6 SEP	VIRGO
6 SEP–30 SEP	LIBRA
30 SEP–25 OCT	SCORPIO
25 OCT–18 NOV	SAGITTARIUS
18 NOV–13 DEC	CAPRICORN
13 DEC–31 DEC	AQUARIUS

1905	Mars in
1 JAN–13 JAN	LIBRA
13 JAN–21 AUG	SCORPIO
21 AUG–8 OCT	SAGITTARIUS
8 OCT–18 NOV	CAPRICORN
18 NOV–27 DEC	AQUARIUS
27 DEC–31 DEC	PISCES

1905	Venus in
1 JAN–7 JAN	AQUARIUS
7 JAN–3 FEB	PISCES
3 FEB–6 MAR	ARIES
6 MAR–9 MAY	TAURUS
9 MAY–28 MAY	ARIES
28 MAY–8 JUL	TAURUS
8 JUL–6 AUG	GEMINI
6 AUG–1 SEP	CANCER
1 SEP–27 SEP	LEO
27 SEP–21 OCT	VIRGO
21 OCT–14 NOV	LIBRA
14 NOV–8 DEC	SCORPIO
8 DEC–31 DEC	SAGITTARIUS

1906	Mars in	1906	Venus in
1 JAN–4 FEB	PISCES	1 JAN	SAGITTARIUS
4 FEB–17 MAR	ARIES	1 JAN–25 JAN	CAPRICORN
[illegible]	[illegible]	[illegible]	[illegible]
		11 AUG–7 SEP	LIBRA
		7 SEP–9 OCT	SCORPIO
		9 OCT–15 DEC	SAGITTARIUS
		15 DEC–25 DEC	SCORPIO
		25 DEC–31 DEC	SAGITTARIUS

1907	Mars in	1907	Venus in
1 JAN–5 FEB	SCORPIO	1 JAN–6 FEB	SAGITTARIUS
5 FEB–1 APR	SAGITTARIUS	6 FEB–6 MAR	CAPRICORN
1 APR–13 OCT	CAPRICORN	6 MAR–2 APR	AQUARIUS
13 OCT–29 NOV	AQUARIUS	2 APR–27 APR	PISCES
29 NOV–31 DEC	PISCES	27 APR–22 MAY	ARIES
		22 MAY–16 JUN	TAURUS
		16 JUN–11 JUL	GEMINI
		11 JUL–4 AUG	CANCER
		4 AUG–29 AUG	LEO
		29 AUG–22 SEP	VIRGO
		22 SEP–16 OCT	LIBRA
		16 OCT–9 NOV	SCORPIO
		9 NOV–3 DEC	SAGITTARIUS
		3 DEC–27 DEC	CAPRICORN
		27 DEC–31 DEC	AQUARIUS

1908	Mars in
1 JAN–11 JAN	PISCES
11 JAN–23 FEB	ARIES
23 FEB–7 APR	TAURUS
7 APR–22 MAY	GEMINI
22 MAY–8 JUL	CANCER
8 JUL–24 AUG	LEO
24 AUG–10 OCT	VIRGO
10 OCT–25 NOV	LIBRA
25 NOV–31 DEC	SCORPIO

1908	Venus in
1 JAN–20 JAN	AQUARIUS
20 JAN–14 FEB	PISCES
14 FEB–10 MAR	ARIES
10 MAR–5 APR	TAURUS
5 APR–5 MAY	GEMINI
5 MAY–8 SEP	CANCER
8 SEP–8 OCT	LEO
8 OCT–3 NOV	VIRGO
3 NOV–28 NOV	LIBRA
28 NOV–22 DEC	SCORPIO
22 DEC–31 DEC	SAGITTARIUS

1909	Mars in
1 JAN–10 JAN	SCORPIO
10 JAN–24 FEB	SAGITTARIUS
24 FEB–9 APR	CAPRICORN
9 APR–25 MAY	AQUARIUS
25 MAY–21 JUL	PISCES
21 JUL–26 SEP	ARIES
26 SEP–20 NOV	PISCES
20 NOV–31 DEC	ARIES

1909	Venus in
1 JAN–15 JAN	SAGITTARIUS
15 JAN–9 FEB	CAPRICORN
9 FEB–5 MAR	AQUARIUS
5 MAR–29 MAR	PISCES
29 MAR–22 APR	ARIES
22 APR–16 MAY	TAURUS
16 MAY–10 JUN	GEMINI
10 JUN–4 JUL	CANCER
4 JUL–29 JUL	LEO
29 JUL–23 AUG	VIRGO
23 AUG–17 SEP	LIBRA
17 SEP–12 OCT	SCORPIO
12 OCT–7 NOV	SAGITTARIUS
7 NOV–5 DEC	CAPRICORN
5 DEC–31 DEC	AQUARIUS

1910	Mars in	1910	Venus in
1 JAN–23 JAN	ARIES	1 JAN–15 JAN	AQUARIUS
		15 JAN–29 JAN	PISCES
		12 SEP–6 OCT	[illegible]
		6 OCT–30 OCT	LIBRA
		30 OCT–23 NOV	SCORPIO
		23 NOV–17 DEC	SAGITTARIUS
		17 DEC–31 DEC	CAPRICORN

1911	Mars in	1911	Venus in
1 JAN–31 JAN	SAGITTARIUS	1 JAN–10 JAN	CAPRICORN
31 JAN–14 MAR	CAPRICORN	10 JAN–3 FEB	AQUARIUS
14 MAR–23 APR	AQUARIUS	3 FEB–27 FEB	PISCES
23 APR–2 JUN	PISCES	27 FEB–23 MAR	ARIES
2 JUN–15 JUL	ARIES	23 MAR–17 APR	TAURUS
15 JUL–5 SEP	TAURUS	17 APR–13 MAY	GEMINI
5 SEP–30 NOV	GEMINI	13 MAY–8 JUN	CANCER
30 NOV–31 DEC	TAURUS	8 JUN–7 JUL	LEO
		7 JUL–9 NOV	VIRGO
		9 NOV–9 DEC	LIBRA
		9 DEC–31 DEC	SCORPIO

1912	Mars in	1912	Venus in
1 JAN–30 JAN	TAURUS	1 JAN–4 JAN	SCORPIO
30 JAN–5 APR	GEMINI	4 JAN–29 JAN	SAGITTARIUS
5 APR–28 MAY	CANCER	29 JAN–23 FEB	CAPRICORN
28 MAY–17 JUL	LEO	23 FEB–19 MAR	AQUARIUS
17 JUL–2 SEP	VIRGO	19 MAR–12 APR	PISCES
2 SEP–18 OCT	LIBRA	12 APR–7 MAY	ARIES
18 OCT–30 NOV	SCORPIO	7 MAY–31 MAY	TAURUS
30 NOV–31 DEC	SAGITTARIUS	31 MAY–25 JUN	GEMINI
		25 JUN–19 JUL	CANCER
		19 JUL–12 AUG	LEO
		12 AUG–6 SEP	VIRGO
		6 SEP–30 SEP	LIBRA
		30 SEP–24 OCT	SCORPIO
		24 OCT–18 NOV	SAGITTARIUS
		18 NOV–12 DEC	CAPRICORN
		12 DEC–31 DEC	AQUARIUS

1913	Mars in	1913	Venus in
1 JAN–10 JAN	SAGITTARIUS	1 JAN–7 JAN	AQUARIUS
10 JAN–19 FEB	CAPRICORN	7 JAN–2 FEB	PISCES
19 FEB–30 MAR	AQUARIUS	2 FEB–6 MAR	ARIES
30 MAR–8 MAY	PISCES	6 MAR–2 MAY	TAURUS
8 MAY–17 JUN	ARIES	2 MAY–31 MAY	ARIES
17 JUN–29 JUL	TAURUS	31 MAY–8 JUL	TAURUS
29 JUL–15 SEP	GEMINI	8 JUL–5 AUG	GEMINI
15 SEP–31 DEC	CANCER	5 AUG–1 SEP	CANCER
		1 SEP–26 SEP	LEO
		26 SEP–21 OCT	VIRGO
		21 OCT–14 NOV	LIBRA
		14 NOV–8 DEC	SCORPIO
		8 DEC–31 DEC	SAGITTARIUS

1914	Mars in
1 JAN–1 MAY	CANCER
[illegible]	[illegible]

1914	Venus in
1 JAN	SAGITTARIUS
[illegible]	CAPRICORN
[illegible]	[illegible]
15 JUL–10 AUG	[illegible]
10 AUG–7 SEP	LIBRA
7 SEP–10 OCT	SCORPIO
10 OCT–5 DEC	SAGITTARIUS
5 DEC–30 DEC	SCORPIO
30 DEC–31 DEC	SAGITTARIUS

1915	Mars in
1 JAN–30 JAN	CAPRICORN
30 JAN–9 MAR	AQUARIUS
9 MAR–16 APR	PISCES
16 APR–26 MAY	ARIES
26 MAY–6 JUL	TAURUS
6 JUL–19 AUG	GEMINI
19 AUG–7 OCT	CANCER
7 OCT–31 DEC	LEO

1915	Venus in
1 JAN–6 FEB	SAGITTARIUS
6 FEB–6 MAR	CAPRICORN
6 MAR–1 APR	AQUARIUS
1 APR–27 APR	PISCES
27 APR–22 MAY	ARIES
22 MAY–16 JUN	TAURUS
16 JUN–10 JUL	GEMINI
10 JUL–4 AUG	CANCER
4 AUG–28 AUG	LEO
28 AUG–21 SEP	VIRGO
21 SEP–15 OCT	LIBRA
15 OCT–8 NOV	SCORPIO
8 NOV–2 DEC	SAGITTARIUS
2 DEC–26 DEC	CAPRICORN
26 DEC–31 DEC	AQUARIUS

1916	Mars in
1 JAN–28 MAY	LEO
28 MAY–23 JUL	VIRGO
23 JUL–8 SEP	LIBRA
8 SEP–22 OCT	SCORPIO
22 OCT–1 DEC	SAGITTARIUS
1 DEC–31 DEC	CAPRICORN

1916	Venus in
1 JAN–20 JAN	AQUARIUS
20 JAN–13 FEB	PISCES
13 FEB–9 MAR	ARIES
9 MAR–5 APR	TAURUS
5 APR–5 MAY	GEMINI
5 MAY–8 SEP	CANCER
8 SEP–7 OCT	LEO
7 OCT–3 NOV	VIRGO
3 NOV–27 NOV	LIBRA
27 NOV–22 DEC	SCORPIO
22 DEC–31 DEC	SAGITTARIUS

1917	Mars in
1 JAN–9 JAN	CAPRICORN
9 JAN–16 FEB	AQUARIUS
16 FEB–26 MAR	PISCES
26 MAR–4 MAY	ARIES
4 MAY–14 JUN	TAURUS
14 JUN–28 JUL	GEMINI
28 JUL–12 SEP	CANCER
12 SEP–2 NOV	LEO
2 NOV–31 DEC	VIRGO

1917	Venus in
1 JAN–15 JAN	SAGITTARIUS
15 JAN–8 FEB	CAPRICORN
8 FEB–4 MAR	AQUARIUS
4 MAR–28 MAR	PISCES
28 MAR–21 APR	ARIES
21 APR–16 MAY	TAURUS
16 MAY–9 JUN	GEMINI
9 JUN–4 JUL	CANCER
4 JUL–28 JUL	LEO
28 JUL–22 AUG	VIRGO
22 AUG–16 SEP	LIBRA
16 SEP–11 OCT	SCORPIO
11 OCT–7 NOV	SAGITTARIUS
7 NOV–5 DEC	CAPRICORN
5 DEC–31 DEC	AQUARIUS

1918	Mars in
[illegible]	VIRGO
[illegible]	[illegible]

1918	Venus in
1 JAN–5 APR	AQUARIUS
[illegible]	PISCES
[illegible]	[illegible]
30 OCT–23 NOV	SCORPIO
23 NOV–17 DEC	SAGITTARIUS
17 DEC–31 DEC	CAPRICORN

1919	Mars in
1 JAN–27 JAN	AQUARIUS
27 JAN–6 MAR	PISCES
6 MAR–15 APR	ARIES
15 APR–26 MAY	TAURUS
26 MAY–8 JUL	GEMINI
8 JUL–23 AUG	CANCER
23 AUG–10 OCT	LEO
10 OCT–30 NOV	VIRGO
30 NOV–31 DEC	LIBRA

1919	Venus in
1 JAN–10 JAN	CAPRICORN
10 JAN–2 FEB	AQUARIUS
2 FEB–27 FEB	PISCES
27 FEB–23 MAR	ARIES
23 MAR–17 APR	TAURUS
17 APR–12 MAY	GEMINI
12 MAY–8 JUN	CANCER
8 JUN–7 JUL	LEO
7 JUL–9 NOV	VIRGO
9 NOV–9 DEC	LIBRA
9 DEC–31 DEC	SCORPIO

1920	Mars in
1 JAN–31 JAN	LIBRA
31 JAN–23 APR	SCORPIO
23 APR–10 JUL	LIBRA
10 JUL–4 SEP	SCORPIO
4 SEP–18 OCT	SAGITTARIUS
18 OCT–27 NOV	CAPRICORN
27 NOV–31 DEC	AQUARIUS

1920	Venus in
1 JAN–4 JAN	SCORPIO
4 JAN–29 JAN	SAGITTARIUS
29 JAN–23 FEB	CAPRICORN
23 FEB–18 MAR	AQUARIUS
18 MAR–12 APR	PISCES
12 APR–6 MAY	ARIES
6 MAY–31 MAY	TAURUS
31 MAY–24 JUN	GEMINI
24 JUN–18 JUL	CANCER
18 JUL–12 AUG	LEO
12 AUG–5 SEP	VIRGO
5 SEP–29 SEP	LIBRA
29 SEP–24 OCT	SCORPIO
24 OCT–17 NOV	SAGITTARIUS
17 NOV–12 DEC	CAPRICORN
12 DEC–31 DEC	AQUARIUS

1921	Mars in
1 JAN–5 JAN	AQUARIUS
5 JAN–13 FEB	PISCES
13 FEB–25 MAR	ARIES
25 MAR–6 MAY	TAURUS
6 MAY–18 JUN	GEMINI
18 JUN–3 AUG	CANCER
3 AUG–19 SEP	LEO
19 SEP–6 NOV	VIRGO
6 NOV–26 DEC	LIBRA
26 DEC–31 DEC	SCORPIO

1921	Venus in
1 JAN–6 JAN	AQUARIUS
6 JAN–2 FEB	PISCES
2 FEB–7 MAR	ARIES
7 MAR–25 APR	TAURUS
25 APR–2 JUN	ARIES
2 JUN–8 JUL	TAURUS
8 JUL–5 AUG	GEMINI
5 AUG–31 AUG	CANCER
31 AUG–26 SEP	LEO
26 SEP–20 OCT	VIRGO
20 OCT–13 NOV	LIBRA
13 NOV–7 DEC	SCORPIO
7 DEC–31 DEC	SAGITTARIUS
31 DEC	CAPRICORN

1922	Mars in
1 JAN–18 FEB	SCORPIO
[illegible]	[illegible]

1922	Venus in
1 JAN–24 JAN	CAPRICORN
24 JAN–[illegible] FEB	AQUARIUS
[illegible]	[illegible]
10 AUG–7 SEP	[illegible]
7 SEP–10 OCT	SCORPIO
10 OCT–28 NOV	SAGITTARIUS
28 NOV–31 DEC	SCORPIO

1923	Mars in
1 JAN–21 JAN	PISCES
21 JAN–4 MAR	ARIES
4 MAR–16 APR	TAURUS
16 APR–30 MAY	GEMINI
30 MAY–16 JUL	CANCER
16 JUL–1 SEP	LEO
1 SEP–18 OCT	VIRGO
18 OCT–4 DEC	LIBRA
4 DEC–31 DEC	SCORPIO

1923	Venus in
1 JAN–2 JAN	SCORPIO
2 JAN–6 FEB	SAGITTARIUS
6 FEB–6 MAR	CAPRICORN
6 MAR–1 APR	AQUARIUS
1 APR–26 APR	PISCES
26 APR–21 MAY	ARIES
21 MAY–15 JUN	TAURUS
15 JUN–10 JUL	GEMINI
10 JUL–3 AUG	CANCER
3 AUG–27 AUG	LEO
27 AUG–21 SEP	VIRGO
21 SEP–15 OCT	LIBRA
15 OCT–8 NOV	SCORPIO
8 NOV–2 DEC	SAGITTARIUS
2 DEC–26 DEC	CAPRICORN
26 DEC–31 DEC	AQUARIUS

1924	Mars in
1 JAN–19 JAN	SCORPIO
19 JAN–6 MAR	SAGITTARIUS
6 MAR–24 APR	CAPRICORN
24 APR–24 JUN	AQUARIUS
24 JUN–24 AUG	PISCES
24 AUG–19 OCT	AQUARIUS
19 OCT–19 DEC	PISCES
19 DEC–31 DEC	ARIES

1924	Venus in
1 JAN–19 JAN	AQUARIUS
19 JAN–13 FEB	PISCES
13 FEB–9 MAR	ARIES
9 MAR–5 APR	TAURUS
5 APR–6 MAY	GEMINI
6 MAY–8 SEP	CANCER
8 SEP–7 OCT	LEO
7 OCT–2 NOV	VIRGO
2 NOV–27 NOV	LIBRA
27 NOV–21 DEC	SCORPIO
21 DEC–31 DEC	SAGITTARIUS

1925	Mars in
1 JAN–5 FEB	ARIES
5 FEB–24 MAR	TAURUS
24 MAR–9 MAY	GEMINI
9 MAY–26 JUN	CANCER
26 JUN–12 AUG	LEO
12 AUG–28 SEP	VIRGO
28 SEP–13 NOV	LIBRA
13 NOV–28 DEC	SCORPIO
28 DEC–31 DEC	SAGITTARIUS

1925	Venus in
1 JAN–14 JAN	SAGITTARIUS
14 JAN–7 FEB	CAPRICORN
7 FEB–4 MAR	AQUARIUS
4 MAR–28 MAR	PISCES
28 MAR–21 APR	ARIES
21 APR–15 MAY	TAURUS
15 MAY–9 JUN	GEMINI
9 JUN–3 JUL	CANCER
3 JUL–28 JUL	LEO
28 JUL–22 AUG	VIRGO
22 AUG–16 SEP	LIBRA
16 SEP–11 OCT	SCORPIO
11 OCT–6 NOV	SAGITTARIUS
6 NOV–5 DEC	CAPRICORN
5 DEC–31 DEC	AQUARIUS

1926	Mars in
1 JAN–9 FEB	SAGITTARIUS

1926	Venus in
1 JAN–6 APR	AQUARIUS
6 APR–6 MAY	PISCES
29 OCT–22 NOV	[illegible]
22 NOV–16 DEC	SAGITTARIUS
16 DEC–31 DEC	CAPRICORN

1927	Mars in
1 JAN–22 FEB	TAURUS
22 FEB–17 APR	GEMINI
17 APR–6 JUN	CANCER
6 JUN–25 JUL	LEO
25 JUL–10 SEP	VIRGO
10 SEP–26 OCT	LIBRA
26 OCT–8 DEC	SCORPIO
8 DEC–31 DEC	SAGITTARIUS

1927	Venus in
1 JAN–9 JAN	CAPRICORN
9 JAN–2 FEB	AQUARIUS
2 FEB–26 FEB	PISCES
26 FEB–22 MAR	ARIES
22 MAR–16 APR	TAURUS
16 APR–12 MAY	GEMINI
12 MAY–8 JUN	CANCER
8 JUN–7 JUL	LEO
7 JUL–9 NOV	VIRGO
9 NOV–8 DEC	LIBRA
8 DEC–31 DEC	SCORPIO

1928	Mars in
1 JAN–19 JAN	SAGITTARIUS
19 JAN–28 FEB	CAPRICORN
28 FEB–7 APR	AQUARIUS
7 APR–16 MAY	PISCES
16 MAY–26 JUN	ARIES
26 JUN–9 AUG	TAURUS
9 AUG–3 OCT	GEMINI
3 OCT–20 DEC	CANCER
20 DEC–31 DEC	GEMINI

1928	Venus in
1 JAN–4 JAN	SCORPIO
4 JAN–29 JAN	SAGITTARIUS
29 JAN–22 FEB	CAPRICORN
22 FEB–18 MAR	AQUARIUS
18 MAR–11 APR	PISCES
11 APR–6 MAY	ARIES
6 MAY–30 MAY	TAURUS
30 MAY–23 JUN	GEMINI
23 JUN–18 JUL	CANCER
18 JUL–11 AUG	LEO
11 AUG–4 SEP	VIRGO
4 SEP–29 SEP	LIBRA
29 SEP–23 OCT	SCORPIO
23 OCT–17 NOV	SAGITTARIUS
17 NOV–12 DEC	CAPRICORN
12 DEC–31 DEC	AQUARIUS

1929	Mars in
1 JAN–10 MAR	GEMINI
10 MAR–13 MAY	CANCER
13 MAY–4 JUL	LEO
4 JUL–21 AUG	VIRGO
21 AUG–6 OCT	LIBRA
6 OCT–18 NOV	SCORPIO
18 NOV–29 DEC	SAGITTARIUS
29 DEC–31 DEC	CAPRICORN

1929	Venus in
1 JAN–6 JAN	AQUARIUS
6 JAN–2 FEB	PISCES
2 FEB–8 MAR	ARIES
8 MAR–20 APR	TAURUS
20 APR–3 JUN	ARIES
3 JUN–8 JUL	TAURUS
8 JUL–5 AUG	GEMINI
5 AUG–31 AUG	CANCER
31 AUG–25 SEP	LEO
25 SEP–20 OCT	VIRGO
20 OCT–13 NOV	LIBRA
13 NOV–7 DEC	SCORPIO
7 DEC–31 DEC	SAGITTARIUS
31 DEC	CAPRICORN

1930	Mars in
1 JAN–6 FEB	CAPRICORN

1930	Venus in
1 JAN–24 JAN	CAPRICORN
24 JAN–16 FEB	AQUARIUS
10 AUG–7 SEP	[illegible]
7 SEP–12 OCT	SCORPIO
12 OCT–22 NOV	SAGITTARIUS
22 NOV–31 DEC	SCORPIO

1931	Mars in
1 JAN–16 FEB	LEO
16 FEB–30 MAR	CANCER
30 MAR–10 JUN	LEO
10 JUN–1 AUG	VIRGO
1 AUG–17 SEP	LIBRA
17 SEP–30 OCT	SCORPIO
30 OCT–10 DEC	SAGITTARIUS
10 DEC–31 DEC	CAPRICORN

1931	Venus in
1 JAN–3 JAN	SCORPIO
3 JAN–6 FEB	SAGITTARIUS
6 FEB–5 MAR	CAPRICORN
5 MAR–31 MAR	AQUARIUS
31 MAR–26 APR	PISCES
26 APR–21 MAY	ARIES
21 MAY–14 JUN	TAURUS
14 JUN–9 JUL	GEMINI
9 JUL–3 AUG	CANCER
3 AUG–27 AUG	LEO
27 AUG–20 SEP	VIRGO
20 SEP–14 OCT	LIBRA
14 OCT–7 NOV	SCORPIO
7 NOV–1 DEC	SAGITTARIUS
1 DEC–25 DEC	CAPRICORN
25 DEC–31 DEC	AQUARIUS

1932	Mars in
1 JAN–18 JAN	CAPRICORN
18 JAN–25 FEB	AQUARIUS
25 FEB–3 APR	PISCES
3 APR–12 MAY	ARIES
12 MAY–22 JUN	TAURUS
22 JUN–4 AUG	GEMINI
4 AUG–20 SEP	CANCER
20 SEP–13 NOV	LEO
13 NOV–31 DEC	VIRGO

1932	Venus in
1 JAN–19 JAN	AQUARIUS
19 JAN–12 FEB	PISCES
12 FEB–9 MAR	ARIES
9 MAR–5 APR	TAURUS
5 APR–6 MAY	GEMINI
6 MAY–13 JUL	CANCER
13 JUL–28 JUL	GEMINI
28 JUL–8 SEP	CANCER
8 SEP–7 OCT	LEO
7 OCT–2 NOV	VIRGO
2 NOV–27 NOV	LIBRA
27 NOV–21 DEC	SCORPIO
21 DEC–31 DEC	SAGITTARIUS

1933	Mars in
1 JAN–6 JUL	VIRGO
6 JUL–26 AUG	LIBRA
26 AUG–9 OCT	SCORPIO
9 OCT–19 NOV	SAGITTARIUS
19 NOV–28 DEC	CAPRICORN
28 DEC–31 DEC	AQUARIUS

1933	Venus in
1 JAN–14 JAN	SAGITTARIUS
14 JAN–7 FEB	CAPRICORN
7 FEB–3 MAR	AQUARIUS
3 MAR–27 MAR	PISCES
27 MAR–20 APR	ARIES
20 APR–15 MAY	TAURUS
15 MAY–8 JUN	GEMINI
8 JUN–3 JUL	CANCER
3 JUL–27 JUL	LEO
27 JUL–21 AUG	VIRGO
21 AUG–15 SEP	LIBRA
15 SEP–11 OCT	SCORPIO
11 OCT–6 NOV	SAGITTARIUS
6 NOV–5 DEC	CAPRICORN
5 DEC–31 DEC	AQUARIUS

1934	**Mars in**	**1934**	**Venus in**
1 JAN–4 FEB	AQUARIUS	1 JAN–6 APR	AQUARIUS
4 FEB–14 MAR	PISCES	6 APR–6 MAY	PISCES
		22 NOV–16 DEC	SAGITTARIUS
		16 DEC–31 DEC	CAPRICORN

1935	**Mars in**	**1935**	**Venus in**
1 JAN–29 JUL	LIBRA	1 JAN–8 JAN	CAPRICORN
29 JUL–16 SEP	SCORPIO	8 JAN–1 FEB	AQUARIUS
16 SEP–28 OCT	SAGITTARIUS	1 FEB–26 FEB	PISCES
28 OCT–7 DEC	CAPRICORN	26 FEB–22 MAR	ARIES
7 DEC–31 DEC	AQUARIUS	22 MAR–16 APR	TAURUS
		16 APR–11 MAY	GEMINI
		11 MAY–7 JUN	CANCER
		7 JUN–7 JUL	LEO
		7 JUL–9 NOV	VIRGO
		9 NOV–8 DEC	LIBRA
		8 DEC–31 DEC	SCORPIO

1936	Mars in
1 JAN–14 JAN	AQUARIUS
14 JAN–22 FEB	PISCES
22 FEB–1 APR	ARIES
1 APR–13 MAY	TAURUS
13 MAY–25 JUN	GEMINI
25 JUN–10 AUG	CANCER
10 AUG–26 SEP	LEO
26 SEP–14 NOV	VIRGO
14 NOV–31 DEC	LIBRA

1936	Venus in
1 JAN–3 JAN	SCORPIO
3 JAN–28 JAN	SAGITTARIUS
28 JAN–22 FEB	CAPRICORN
22 FEB–17 MAR	AQUARIUS
17 MAR–11 APR	PISCES
11 APR–5 MAY	ARIES
5 MAY–29 MAY	TAURUS
29 MAY–23 JUN	GEMINI
23 JUN–17 JUL	CANCER
17 JUL–11 AUG	LEO
11 AUG–4 SEP	VIRGO
4 SEP–28 SEP	LIBRA
28 SEP–23 OCT	SCORPIO
23 OCT–16 NOV	SAGITTARIUS
16 NOV–11 DEC	CAPRICORN
11 DEC–31 DEC	AQUARIUS

1937	Mars in
1 JAN–5 JAN	LIBRA
5 JAN–13 MAR	SCORPIO
13 MAR–14 MAY	SAGITTARIUS
14 MAY–8 AUG	SCORPIO
8 AUG–30 SEP	SAGITTARIUS
30 SEP–11 NOV	CAPRICORN
11 NOV–21 DEC	AQUARIUS
21 DEC–31 DEC	PISCES

1937	Venus in
1 JAN–6 JAN	AQUARIUS
6 JAN–2 FEB	PISCES
2 FEB–9 MAR	ARIES
9 MAR–14 APR	TAURUS
14 APR–4 JUN	ARIES
4 JUN–7 JUL	TAURUS
7 JUL–4 AUG	GEMINI
4 AUG–31 AUG	CANCER
31 AUG–25 SEP	LEO
25 SEP–19 OCT	VIRGO
19 OCT–12 NOV	LIBRA
12 NOV–6 DEC	SCORPIO
6 DEC–30 DEC	SAGITTARIUS
30 DEC–31 DEC	CAPRICORN

1938	Mars in
1 JAN–30 JAN	PISCES
[illegible]	[illegible]

1938	Venus in
1 JAN–23 JAN	CAPRICORN
[illegible]	AQUARIUS
[illegible]	[illegible]
9 AUG–7 SEP	LIBRA
7 SEP–13 OCT	SCORPIO
13 OCT–15 NOV	SAGITTARIUS
15 NOV–31 DEC	SCORPIO

1939	Mars in
1 JAN–29 JAN	SCORPIO
29 JAN–21 MAR	SAGITTARIUS
21 MAR–25 MAY	CAPRICORN
25 MAY–21 JUL	AQUARIUS
21 JUL–24 SEP	CAPRICORN
24 SEP–19 NOV	AQUARIUS
19 NOV–31 DEC	PISCES

1939	Venus in
1 JAN–4 JAN	SCORPIO
4 JAN–6 FEB	SAGITTARIUS
6 FEB–5 MAR	CAPRICORN
5 MAR–31 MAR	AQUARIUS
31 MAR–25 APR	PISCES
25 APR–20 MAY	ARIES
20 MAY–14 JUN	TAURUS
14 JUN–9 JUL	GEMINI
9 JUL–2 AUG	CANCER
2 AUG–26 AUG	LEO
26 AUG–20 SEP	VIRGO
20 SEP–14 OCT	LIBRA
14 OCT–7 NOV	SCORPIO
7 NOV–1 DEC	SAGITTARIUS
1 DEC–25 DEC	CAPRICORN
25 DEC–31 DEC	AQUARIUS

1940	Mars in
1 JAN–4 JAN	PISCES
4 JAN–17 FEB	ARIES
17 FEB–1 APR	TAURUS
1 APR–17 MAY	GEMINI
17 MAY–3 JUL	CANCER
3 JUL–19 AUG	LEO
19 AUG–5 OCT	VIRGO
5 OCT–20 NOV	LIBRA
20 NOV–31 DEC	SCORPIO

1940	Venus in
1 JAN–18 JAN	AQUARIUS
18 JAN–12 FEB	PISCES
12 FEB–8 MAR	ARIES
8 MAR–4 APR	TAURUS
4 APR–6 MAY	GEMINI
6 MAY–5 JUL	CANCER
5 JUL–1 AUG	GEMINI
1 AUG–8 SEP	CANCER
8 SEP–6 OCT	LEO
6 OCT–1 NOV	VIRGO
1 NOV–26 NOV	LIBRA
26 NOV–20 DEC	SCORPIO
20 DEC–31 DEC	SAGITTARIUS

1941	Mars in
1 JAN–4 JAN	SCORPIO
4 JAN–17 FEB	SAGITTARIUS
17 FEB–2 APR	CAPRICORN
2 APR–16 MAY	AQUARIUS
16 MAY–2 JUL	PISCES
2 JUL–31 DEC	ARIES

1941	Venus in
1 JAN–13 JAN	SAGITTARIUS
13 JAN–6 FEB	CAPRICORN
6 FEB–2 MAR	AQUARIUS
2 MAR–27 MAR	PISCES
27 MAR–20 APR	ARIES
20 APR–14 MAY	TAURUS
14 MAY–7 JUN	GEMINI
7 JUN–2 JUL	CANCER
2 JUL–27 JUL	LEO
27 JUL–21 AUG	VIRGO
21 AUG–15 SEP	LIBRA
15 SEP–10 OCT	SCORPIO
10 OCT–6 NOV	SAGITTARIUS
6 NOV–5 DEC	CAPRICORN
5 DEC–31 DEC	AQUARIUS

1942	Mars in
[illegible]	[illegible]
1 NOV [illegible]	[illegible]
15 DEC–31 DEC	SAGITTARIUS

1942	Venus in
1 JAN–6 APR	AQUARIUS
[illegible]	[illegible]
28 OCT–21 NOV	SCORPIO
21 NOV–15 DEC	SAGITTARIUS
15 DEC–31 DEC	CAPRICORN

1943	Mars in
1 JAN–26 JAN	SAGITTARIUS
26 JAN–8 MAR	CAPRICORN
8 MAR–17 APR	AQUARIUS
17 APR–27 MAY	PISCES
27 MAY–7 JUL	ARIES
7 JUL–23 AUG	TAURUS
23 AUG–31 DEC	GEMINI

1943	Venus in
1 JAN–8 JAN	CAPRICORN
8 JAN–1 FEB	AQUARIUS
1 FEB–25 FEB	PISCES
25 FEB–21 MAR	ARIES
21 MAR–15 APR	TAURUS
15 APR–11 MAY	GEMINI
11 MAY–7 JUN	CANCER
7 JUN–7 JUL	LEO
7 JUL–9 NOV	VIRGO
9 NOV–8 DEC	LIBRA
8 DEC–31 DEC	SCORPIO

1944	Mars in
1 JAN–28 MAR	GEMINI
28 MAR–22 MAY	CANCER
22 MAY–12 JUL	LEO
12 JUL–29 AUG	VIRGO
29 AUG–13 OCT	LIBRA
13 OCT–25 NOV	SCORPIO
25 NOV–31 DEC	SAGITTARIUS

1944	Venus in
1 JAN–3 JAN	SCORPIO
3 JAN–28 JAN	SAGITTARIUS
28 JAN–21 FEB	CAPRICORN
21 FEB–17 MAR	AQUARIUS
17 MAR–10 APR	PISCES
10 APR–4 MAY	ARIES
4 MAY–29 MAY	TAURUS
29 MAY–22 JUN	GEMINI
22 JUN–17 JUL	CANCER
17 JUL–10 AUG	LEO
10 AUG–3 SEP	VIRGO
3 SEP–28 SEP	LIBRA
28 SEP–22 OCT	SCORPIO
22 OCT–16 NOV	SAGITTARIUS
16 NOV–11 DEC	CAPRICORN
11 DEC–31 DEC	AQUARIUS

1945	Mars in
1 JAN–5 JAN	SAGITTARIUS
5 JAN–14 FEB	CAPRICORN
14 FEB–25 MAR	AQUARIUS
25 MAR–2 MAY	PISCES
2 MAY–11 JUN	ARIES
11 JUN–23 JUL	TAURUS
23 JUL–7 SEP	GEMINI
7 SEP–11 NOV	CANCER
11 NOV–26 DEC	LEO
26 DEC–31 DEC	CANCER

1945	Venus in
1 JAN–5 JAN	AQUARIUS
5 JAN–2 FEB	PISCES
2 FEB–11 MAR	ARIES
11 MAR–7 APR	TAURUS
7 APR–4 JUN	ARIES
4 JUN–7 JUL	TAURUS
7 JUL–4 AUG	GEMINI
4 AUG–30 AUG	CANCER
30 AUG–24 SEP	LEO
24 SEP–19 OCT	VIRGO
19 OCT–12 NOV	LIBRA
12 NOV–6 DEC	SCORPIO
6 DEC–30 DEC	SAGITTARIUS
30 DEC–31 DEC	CAPRICORN

1946	Mars in	1946	Venus in
		1 JAN–22 JAN	CAPRICORN
		[illegible]	[illegible]
		9 AUG–7 SEP	LIBRA
		7 SEP–16 OCT	SCORPIO
		16 OCT–8 NOV	SAGITTARIUS
		8 NOV–31 DEC	SCORPIO

1947	Mars in	1947	Venus in
1 JAN–25 JAN	CAPRICORN	1 JAN–5 JAN	SCORPIO
25 JAN–4 MAR	AQUARIUS	5 JAN–6 FEB	SAGITTARIUS
4 MAR–11 APR	PISCES	6 FEB–5 MAR	CAPRICORN
11 APR–21 MAY	ARIES	5 MAR–30 MAR	AQUARIUS
21 MAY–1 JUL	TAURUS	30 MAR–25 APR	PISCES
1 JUL–13 AUG	GEMINI	25 APR–20 MAY	ARIES
13 AUG–1 OCT	CANCER	20 MAY–13 JUN	TAURUS
1 OCT–1 DEC	LEO	13 JUN–8 JUL	GEMINI
1 DEC–31 DEC	VIRGO	8 JUL–2 AUG	CANCER
		2 AUG–26 AUG	LEO
		26 AUG–19 SEP	VIRGO
		19 SEP–13 OCT	LIBRA
		13 OCT–6 NOV	SCORPIO
		6 NOV–30 NOV	SAGITTARIUS
		30 NOV–24 DEC	CAPRICORN
		24 DEC–31 DEC	AQUARIUS

1948	Mars in
1 JAN–12 FEB	VIRGO
12 FEB–18 MAY	LEO
18 MAY–17 JUL	VIRGO
17 JUL–3 SEP	LIBRA
3 SEP–17 OCT	SCORPIO
17 OCT–26 NOV	SAGITTARIUS
26 NOV–31 DEC	CAPRICORN

1948	Venus in
1 JAN–18 JAN	AQUARIUS
18 JAN–11 FEB	PISCES
11 FEB–8 MAR	ARIES
8 MAR–4 APR	TAURUS
4 APR–7 MAY	GEMINI
7 MAY–29 JUN	CANCER
29 JUN–3 AUG	GEMINI
3 AUG–8 SEP	CANCER
8 SEP–6 OCT	LEO
6 OCT–1 NOV	VIRGO
1 NOV–26 NOV	LIBRA
26 NOV–20 DEC	SCORPIO
20 DEC–31 DEC	SAGITTARIUS

1949	Mars in
1 JAN–4 JAN	CAPRICORN
4 JAN–11 FEB	AQUARIUS
11 FEB–21 MAR	PISCES
21 MAR–30 APR	ARIES
30 APR–10 JUN	TAURUS
10 JUN–23 JUL	GEMINI
23 JUL–7 SEP	CANCER
7 SEP–27 OCT	LEO
27 OCT–26 DEC	VIRGO
26 DEC–31 DEC	LIBRA

1949	Venus in
1 JAN–13 JAN	SAGITTARIUS
13 JAN–6 FEB	CAPRICORN
6 FEB–2 MAR	AQUARIUS
2 MAR–26 MAR	PISCES
26 MAR–19 APR	ARIES
19 APR–14 MAY	TAURUS
14 MAY–7 JUN	GEMINI
7 JUN–1 JUL	CANCER
1 JUL–26 JUL	LEO
26 JUL–20 AUG	VIRGO
20 AUG–14 SEP	LIBRA
14 SEP–10 OCT	SCORPIO
10 OCT–6 NOV	SAGITTARIUS
6 NOV–6 DEC	CAPRICORN
6 DEC–31 DEC	AQUARIUS

1950	Mars in	1950	Venus in
		1 JAN–6 APR	AQUARIUS
[illegible]	[illegible]	[illegible]	[illegible]
		28 OCT–21 NOV	SCORPIO
		21 NOV–14 DEC	SAGITTARIUS
		14 DEC–31 DEC	CAPRICORN

1951	Mars in	1951	Venus in
1 JAN–22 JAN	AQUARIUS	1 JAN–7 JAN	CAPRICORN
22 JAN–1 MAR	PISCES	7 JAN–31 JAN	AQUARIUS
1 MAR–10 APR	ARIES	31 JAN–24 FEB	PISCES
10 APR–21 MAY	TAURUS	24 FEB–21 MAR	ARIES
21 MAY–3 JUL	GEMINI	21 MAR–15 APR	TAURUS
3 JUL–18 AUG	CANCER	15 APR–11 MAY	GEMINI
18 AUG–5 OCT	LEO	11 MAY–7 JUN	CANCER
5 OCT–24 NOV	VIRGO	7 JUN–8 JUL	LEO
24 NOV–31 DEC	LIBRA	8 JUL–9 NOV	VIRGO
		9 NOV–8 DEC	LIBRA
		8 DEC–31 DEC	SCORPIO

1952	Mars in
1 JAN–20 JAN	LIBRA
20 JAN–27 AUG	SCORPIO
27 AUG–12 OCT	SAGITTARIUS
12 OCT–21 NOV	CAPRICORN
21 NOV–30 DEC	AQUARIUS
30 DEC–31 DEC	PISCES

1952	Venus in
1 JAN–2 JAN	SCORPIO
2 JAN–27 JAN	SAGITTARIUS
27 JAN–21 FEB	CAPRICORN
21 FEB–16 MAR	AQUARIUS
16 MAR–9 APR	PISCES
9 APR–4 MAY	ARIES
4 MAY–28 MAY	TAURUS
28 MAY–22 JUN	GEMINI
22 JUN–16 JUL	CANCER
16 JUL–9 AUG	LEO
9 AUG–3 SEP	VIRGO
3 SEP–27 SEP	LIBRA
27 SEP–22 OCT	SCORPIO
22 OCT–15 NOV	SAGITTARIUS
15 NOV–10 DEC	CAPRICORN
10 DEC–31 DEC	AQUARIUS

1953	Mars in
1 JAN–8 FEB	PISCES
8 FEB–20 MAR	ARIES
20 MAR–1 MAY	TAURUS
1 MAY–14 JUN	GEMINI
14 JUN–29 JUL	CANCER
29 JUL–14 SEP	LEO
14 SEP–1 NOV	VIRGO
1 NOV–20 DEC	LIBRA
20 DEC–31 DEC	SCORPIO

1953	Venus in
1 JAN–5 JAN	AQUARIUS
5 JAN–2 FEB	PISCES
2 FEB–14 MAR	ARIES
14 MAR–31 MAR	TAURUS
31 MAR–5 JUN	ARIES
5 JUN–7 JUL	TAURUS
7 JUL–4 AUG	GEMINI
4 AUG–30 AUG	CANCER
30 AUG–24 SEP	LEO
24 SEP–18 OCT	VIRGO
18 OCT–11 NOV	LIBRA
11 NOV–5 DEC	SCORPIO
5 DEC–29 DEC	SAGITTARIUS
29 DEC–31 DEC	CAPRICORN

1954	Mars in
[illegible]	SCORPIO

1954	Venus in
1 JAN–22 JAN	CAPRICORN
[illegible]	AQUARIUS
9 AUG–6 SEP	LIBRA
6 SEP–23 OCT	SCORPIO
23 OCT–27 OCT	SAGITTARIUS
27 OCT–31 DEC	SCORPIO

1955	Mars in
1 JAN–15 JAN	PISCES
15 JAN–26 FEB	ARIES
26 FEB–10 APR	TAURUS
10 APR–26 MAY	GEMINI
26 MAY–11 JUL	CANCER
11 JUL–27 AUG	LEO
27 AUG–13 OCT	VIRGO
13 OCT–29 NOV	LIBRA
29 NOV–31 DEC	SCORPIO

1955	Venus in
1 JAN–6 JAN	SCORPIO
6 JAN–6 FEB	SAGITTARIUS
6 FEB–4 MAR	CAPRICORN
4 MAR–30 MAR	AQUARIUS
30 MAR–24 APR	PISCES
24 APR–19 MAY	ARIES
19 MAY–13 JUN	TAURUS
13 JUN–8 JUL	GEMINI
8 JUL–1 AUG	CANCER
1 AUG–25 AUG	LEO
25 AUG–18 SEP	VIRGO
18 SEP–13 OCT	LIBRA
13 OCT–6 NOV	SCORPIO
6 NOV–30 NOV	SAGITTARIUS
30 NOV–24 DEC	CAPRICORN
24 DEC–31 DEC	AQUARIUS

1956	Mars in
1 JAN–14 JAN	SCORPIO
14 JAN–28 FEB	SAGITTARIUS
28 FEB–14 APR	CAPRICORN
14 APR–3 JUN	AQUARIUS
3 JUN–6 DEC	PISCES
6 DEC–31 DEC	ARIES

1956	Venus in
1 JAN–17 JAN	AQUARIUS
17 JAN–11 FEB	PISCES
11 FEB–7 MAR	ARIES
7 MAR–4 APR	TAURUS
4 APR–8 MAY	GEMINI
8 MAY–23 JUN	CANCER
23 JUN–4 AUG	GEMINI
4 AUG–8 SEP	CANCER
8 SEP–6 OCT	LEO
6 OCT–31 OCT	VIRGO
31 OCT–25 NOV	LIBRA
25 NOV–19 DEC	SCORPIO
19 DEC–31 DEC	SAGITTARIUS

1957	Mars in
1 JAN–28 JAN	ARIES
28 JAN–17 MAR	TAURUS
17 MAR–4 MAY	GEMINI
4 MAY–21 JUN	CANCER
21 JUN–8 AUG	LEO
8 AUG–24 SEP	VIRGO
24 SEP–8 NOV	LIBRA
8 NOV–23 DEC	SCORPIO
23 DEC–31 DEC	SAGITTARIUS

1957	Venus in
1 JAN–12 JAN	SAGITTARIUS
12 JAN–5 FEB	CAPRICORN
5 FEB–1 MAR	AQUARIUS
1 MAR–25 MAR	PISCES
25 MAR–19 APR	ARIES
19 APR–13 MAY	TAURUS
13 MAY–6 JUN	GEMINI
6 JUN–1 JUL	CANCER
1 JUL–26 JUL	LEO
26 JUL–20 AUG	VIRGO
20 AUG–14 SEP	LIBRA
14 SEP–10 OCT	SCORPIO
10 OCT–5 NOV	SAGITTARIUS
5 NOV–6 DEC	CAPRICORN
6 DEC–31 DEC	AQUARIUS

1958	Mars in
[illegible]	SAGITTARIUS
[illegible]	[illegible]

1958	Venus in
1 JAN–6 APR	AQUARIUS
[illegible]	PISCES
[illegible]	[illegible]
27 OCT–20 NOV	SCORPIO
20 NOV–14 DEC	SAGITTARIUS
14 DEC–31 DEC	CAPRICORN

1959	Mars in
1 JAN–10 FEB	TAURUS
10 FEB–10 APR	GEMINI
10 APR–1 JUN	CANCER
1 JUN–20 JUL	LEO
20 JUL–5 SEP	VIRGO
5 SEP–21 OCT	LIBRA
21 OCT–3 DEC	SCORPIO
3 DEC–31 DEC	SAGITTARIUS

1959	Venus in
1 JAN–7 JAN	CAPRICORN
7 JAN–31 JAN	AQUARIUS
31 JAN–24 FEB	PISCES
24 FEB–20 MAR	ARIES
20 MAR–14 APR	TAURUS
14 APR–10 MAY	GEMINI
10 MAY–6 JUN	CANCER
6 JUN–8 JUL	LEO
8 JUL–20 SEP	VIRGO
20 SEP–25 SEP	LEO
25 SEP–9 NOV	VIRGO
9 NOV–7 DEC	LIBRA
7 DEC–31 DEC	SCORPIO

1960	Mars in
1 JAN–14 JAN	SAGITTARIUS
14 JAN–23 FEB	CAPRICORN
23 FEB–2 APR	AQUARIUS
2 APR–11 MAY	PISCES
11 MAY–20 JUN	ARIES
20 JUN–2 AUG	TAURUS
2 AUG–21 SEP	GEMINI
21 SEP–31 DEC	CANCER

1960	Venus in
1 JAN–2 JAN	SCORPIO
2 JAN–27 JAN	SAGITTARIUS
27 JAN–20 FEB	CAPRICORN
20 FEB–16 MAR	AQUARIUS
16 MAR–9 APR	PISCES
9 APR–3 MAY	ARIES
3 MAY–28 MAY	TAURUS
28 MAY–21 JUN	GEMINI
21 JUN–16 JUL	CANCER
16 JUL–9 AUG	LEO
9 AUG–2 SEP	VIRGO
2 SEP–27 SEP	LIBRA
27 SEP–21 OCT	SCORPIO
21 OCT–15 NOV	SAGITTARIUS
15 NOV–10 DEC	CAPRICORN
10 DEC–31 DEC	AQUARIUS

1961	Mars in
1 JAN–5 FEB	CANCER
5 FEB–7 FEB	GEMINI
7 FEB–6 MAY	CANCER
6 MAY–28 JUN	LEO
28 JUN–17 AUG	VIRGO
17 AUG–1 OCT	LIBRA
1 OCT–13 NOV	SCORPIO
13 NOV–24 DEC	SAGITTARIUS
24 DEC–31 DEC	CAPRICORN

1961	Venus in
1 JAN–5 JAN	AQUARIUS
5 JAN–2 FEB	PISCES
2 FEB–5 JUN	ARIES
5 JUN–7 JUL	TAURUS
7 JUL–3 AUG	GEMINI
3 AUG–29 AUG	CANCER
29 AUG–23 SEP	LEO
23 SEP–18 OCT	VIRGO
18 OCT–11 NOV	LIBRA
11 NOV–5 DEC	SCORPIO
5 DEC–29 DEC	SAGITTARIUS
29 DEC–31 DEC	CAPRICORN

1962	Mars in	1962	Venus in
[illegible]	CAPRICORN	1 JAN–21 JAN	CAPRICORN
		[illegible]	AQUARIUS
		[illegible]	[illegible]
		8 AUG–7 SEP	[illegible]
		7 SEP–31 DEC	SCORPIO

1963	Mars in	1963	Venus in
1 JAN–3 JUN	LEO	1 JAN–6 JAN	SCORPIO
3 JUN–27 JUL	VIRGO	6 JAN–5 FEB	SAGITTARIUS
27 JUL–12 SEP	LIBRA	5 FEB–4 MAR	CAPRICORN
12 SEP–25 OCT	SCORPIO	4 MAR–30 MAR	AQUARIUS
25 OCT–5 DEC	SAGITTARIUS	30 MAR–24 APR	PISCES
5 DEC–31 DEC	CAPRICORN	24 APR–19 MAY	ARIES
		19 MAY–12 JUN	TAURUS
		12 JUN–7 JUL	GEMINI
		7 JUL–31 JUL	CANCER
		31 JUL–25 AUG	LEO
		25 AUG–18 SEP	VIRGO
		18 SEP–12 OCT	LIBRA
		12 OCT–5 NOV	SCORPIO
		5 NOV–29 NOV	SAGITTARIUS
		29 NOV–23 DEC	CAPRICORN
		23 DEC–31 DEC	AQUARIUS

1964	Mars in
1 JAN–13 JAN	CAPRICORN
13 JAN–20 FEB	AQUARIUS
20 FEB–29 MAR	PISCES
29 MAR–7 MAY	ARIES
7 MAY–17 JUN	TAURUS
17 JUN–30 JUL	GEMINI
30 JUL–15 SEP	CANCER
15 SEP–6 NOV	LEO
6 NOV–31 DEC	VIRGO

1964	Venus in
1 JAN–17 JAN	AQUARIUS
17 JAN–10 FEB	PISCES
10 FEB–7 MAR	ARIES
7 MAR–4 APR	TAURUS
4 APR–9 MAY	GEMINI
9 MAY–17 JUN	CANCER
17 JUN–5 AUG	GEMINI
5 AUG–8 SEP	CANCER
8 SEP–5 OCT	LEO
5 OCT–31 OCT	VIRGO
31 OCT–25 NOV	LIBRA
25 NOV–19 DEC	SCORPIO
19 DEC–31 DEC	SAGITTARIUS

1965	Mars in
1 JAN–29 JUN	VIRGO
29 JUN–20 AUG	LIBRA
20 AUG–4 OCT	SCORPIO
4 OCT–14 NOV	SAGITTARIUS
14 NOV–23 DEC	CAPRICORN
23 DEC–31 DEC	AQUARIUS

1965	Venus in
1 JAN–12 JAN	SAGITTARIUS
12 JAN–5 FEB	CAPRICORN
5 FEB–1 MAR	AQUARIUS
1 MAR–25 MAR	PISCES
25 MAR–18 APR	ARIES
18 APR–12 MAY	TAURUS
12 MAY–6 JUN	GEMINI
6 JUN–30 JUN	CANCER
30 JUN–25 JUL	LEO
25 JUL–19 AUG	VIRGO
19 AUG–13 SEP	LIBRA
13 SEP–9 OCT	SCORPIO
9 OCT–5 NOV	SAGITTARIUS
5 NOV–7 DEC	CAPRICORN
7 DEC–31 DEC	AQUARIUS

[illegible]	Mars in
[illegible]	[illegible]
12 OCT–4 DEC	V[illegible]
4 DEC–31 DEC	LIBRA

1966	Venus in
[illegible]AN–6 FEB	AQUARIUS
[illegible]	[illegible]PRICORN
[illegible]	[illegible]
13 A[illegible]	[illegible]
8 SEP–3 OCT	VIRG[illegible]
3 OCT–27 OCT	LIBRA
27 OCT–20 NOV	SCORPIO
20 NOV–13 DEC	SAGITTARIUS
13 DEC–31 DEC	CAPRICORN

1967	Mars in
1 JAN–12 FEB	LIBRA
12 FEB–31 MAR	SCORPIO
31 MAR–19 JUL	LIBRA
19 JUL–10 SEP	SCORPIO
10 SEP–23 OCT	SAGITTARIUS
23 OCT–1 DEC	CAPRICORN
1 DEC–31 DEC	AQUARIUS

1967	Venus in
1 JAN–6 JAN	CAPRICORN
6 JAN–30 JAN	AQUARIUS
30 JAN–23 FEB	PISCES
23 FEB–20 MAR	ARIES
20 MAR–14 APR	TAURUS
14 APR–10 MAY	GEMINI
10 MAY–6 JUN	CANCER
6 JUN–8 JUL	LEO
8 JUL–9 SEP	VIRGO
9 SEP–1 OCT	LEO
1 OCT–9 NOV	VIRGO
9 NOV–7 DEC	LIBRA
7 DEC–31 DEC	SCORPIO

1968	Mars in
1 JAN–9 JAN	AQUARIUS
9 JAN–17 FEB	PISCES
17 FEB–27 MAR	ARIES
27 MAR–8 MAY	TAURUS
8 MAY–21 JUN	GEMINI
21 JUN–5 AUG	CANCER
5 AUG–21 SEP	LEO
21 SEP–9 NOV	VIRGO
9 NOV–29 DEC	LIBRA
29 DEC–31 DEC	SCORPIO

1968	Venus in
1 JAN	SCORPIO
1 JAN–26 JAN	SAGITTARIUS
26 JAN–20 FEB	CAPRICORN
20 FEB–15 MAR	AQUARIUS
15 MAR–8 APR	PISCES
8 APR–3 MAY	ARIES
3 MAY–27 MAY	TAURUS
27 MAY–21 JUN	GEMINI
21 JUN–15 JUL	CANCER
15 JUL–8 AUG	LEO
8 AUG–2 SEP	VIRGO
2 SEP–26 SEP	LIBRA
26 SEP–21 OCT	SCORPIO
21 OCT–14 NOV	LIBRA
14 NOV–9 DEC	CAPRICORN
9 DEC–31 DEC	AQUARIUS

1969	Mars in
1 JAN–25 FEB	SCORPIO
25 FEB–21 SEP	SAGITTARIUS
21 SEP–4 NOV	CAPRICORN
4 NOV–15 DEC	AQUARIUS
15 DEC–31 DEC	PISCES

1969	Venus in
1 JAN–4 JAN	AQUARIUS
4 JAN–2 FEB	PISCES
2 FEB–6 JUN	ARIES
6 JUN–6 JUL	TAURUS
6 JUL–3 AUG	GEMINI
3 AUG–29 AUG	CANCER
29 AUG–23 SEP	LEO
23 SEP–17 OCT	VIRGO
17 OCT–10 NOV	LIBRA
10 NOV–4 DEC	SCORPIO
4 DEC–28 DEC	SAGITTARIUS
28 DEC–31 DEC	CAPRICORN

1970	Mars in
[illegible]	[illegible]
20 OCT–6 DEC	[illegible]
6 DEC–31 DEC	SCORPIO

1970	Venus in
1 JAN–21 JAN	CAPRICORN
[illegible]	AQUARIUS
[illegible]	[illegible]
8 AUG–7 SEP	LIBRA
7 SEP–31 DEC	SCORPIO

1971	Mars in
1 JAN–23 JAN	SCORPIO
23 JAN–12 MAR	SAGITTARIUS
12 MAR–3 MAY	CAPRICORN
3 MAY–6 NOV	AQUARIUS
6 NOV–26 DEC	PISCES
26 DEC–31 DEC	ARIES

1971	Venus in
1 JAN–7 JAN	SCORPIO
7 JAN–5 FEB	SAGITTARIUS
5 FEB–4 MAR	CAPRICORN
4 MAR–29 MAR	AQUARIUS
29 MAR–23 APR	PISCES
23 APR–18 MAY	ARIES
18 MAY–12 JUN	TAURUS
12 JUN–6 JUL	GEMINI
6 JUL–31 JUL	CANCER
31 JUL–24 AUG	LEO
24 AUG–17 SEP	VIRGO
17 SEP–11 OCT	LIBRA
11 OCT–5 NOV	SCORPIO
5 NOV–29 NOV	SAGITTARIUS
29 NOV–23 DEC	CAPRICORN
23 DEC–31 DEC	AQUARIUS

1972	Mars in
1 JAN–10 FEB	ARIES
10 FEB–27 MAR	TAURUS
27 MAR–12 MAY	GEMINI
12 MAY–28 JUN	CANCER
28 JUN–15 AUG	LEO
15 AUG–30 SEP	VIRGO
30 SEP–15 NOV	LIBRA
15 NOV–30 DEC	SCORPIO
30 DEC–31 DEC	SAGITTARIUS

1972	Venus in
1 JAN–16 JAN	AQUARIUS
16 JAN–10 FEB	PISCES
10 FEB–7 MAR	ARIES
7 MAR–3 APR	TAURUS
3 APR–10 MAY	GEMINI
10 MAY–11 JUN	CANCER
11 JUN–6 AUG	GEMINI
6 AUG–7 SEP	CANCER
7 SEP–5 OCT	LEO
5 OCT–30 OCT	VIRGO
30 OCT–24 NOV	LIBRA
24 NOV–18 DEC	SCORPIO
18 DEC–31 DEC	SAGITTARIUS

1973	Mars in
1 JAN–12 FEB	SAGITTARIUS
12 FEB–26 MAR	CAPRICORN
26 MAR–8 MAY	AQUARIUS
8 MAY–20 JUN	PISCES
20 JUN–12 AUG	ARIES
12 AUG–29 OCT	TAURUS
29 OCT–24 DEC	ARIES
24 DEC–31 DEC	TAURUS

1973	Venus in
1 JAN–11 JAN	SAGITTARIUS
11 JAN–4 FEB	CAPRICORN
4 FEB–28 FEB	AQUARIUS
28 FEB–24 MAR	PISCES
24 MAR–18 APR	ARIES
18 APR–12 MAY	TAURUS
12 MAY–5 JUN	GEMINI
5 JUN–30 JUN	CANCER
30 JUN–25 JUL	LEO
25 JUL–19 AUG	VIRGO
19 AUG–13 SEP	LIBRA
13 SEP–9 OCT	SCORPIO
9 OCT–5 NOV	SAGITTARIUS
5 NOV–7 DEC	CAPRICORN
7 DEC–31 DEC	AQUARIUS

1974	Mars in	1974	Venus in
1 JAN–27 FEB	TAURUS	1 JAN–29 JAN	AQUARIUS
[illegible]	[illegible]	29 JAN–28 FEB	CAPRICORN
[illegible]	[illegible]	[illegible]	[illegible]
		[illegible]	[illegible]
		2 OCT–26 OCT	LIBRA
		26 OCT–19 NOV	SCORPIO
		19 NOV–13 DEC	SAGITTARIUS
		13 DEC–31 DEC	CAPRICORN

1975	Mars in	1975	Venus in
1 JAN–21 JAN	SAGITTARIUS	1 JAN–6 JAN	CAPRICORN
21 JAN–3 MAR	CAPRICORN	6 JAN–30 JAN	AQUARIUS
3 MAR–11 APR	AQUARIUS	30 JAN–23 FEB	PISCES
11 APR–21 MAY	PISCES	23 FEB–19 MAR	ARIES
21 MAY–1 JUL	ARIES	19 MAR–13 APR	TAURUS
1 JUL–14 AUG	TAURUS	13 APR–9 MAY	GEMINI
14 AUG–17 OCT	GEMINI	9 MAY–6 JUN	CANCER
17 OCT–25 NOV	CANCER	6 JUN–9 JUL	LEO
25 NOV–31 DEC	GEMINI	9 JUL–2 SEP	VIRGO
		2 SEP–4 OCT	LEO
		4 OCT–9 NOV	VIRGO
		9 NOV–7 DEC	LIBRA
		7 DEC–31 DEC	SCORPIO

1976	Mars in
1 JAN–18 MAR	GEMINI
18 MAR–16 MAY	CANCER
16 MAY–6 JUL	LEO
6 JUL–24 AUG	VIRGO
24 AUG–8 OCT	LIBRA
8 OCT–20 NOV	SCORPIO
20 NOV–31 DEC	SAGITTARIUS

1976	Venus in
1 JAN	SCORPIO
1 JAN–26 JAN	SAGITTARIUS
26 JAN–19 FEB	CAPRICORN
19 FEB–15 MAR	AQUARIUS
15 MAR–8 APR	PISCES
8 APR–2 MAY	ARIES
2 MAY–27 MAY	TAURUS
27 MAY–20 JUN	GEMINI
20 JUN–14 JUL	CANCER
14 JUL–8 AUG	LEO
8 AUG–1 SEP	VIRGO
1 SEP–26 SEP	LIBRA
26 SEP–20 OCT	SCORPIO
20 OCT–14 NOV	SAGITTARIUS
14 NOV–9 DEC	CAPRICORN
9 DEC–31 DEC	AQUARIUS

1977	Mars in
1 JAN	SAGITTARIUS
1 JAN–9 FEB	CAPRICORN
9 FEB–20 MAR	AQUARIUS
20 MAR–27 APR	PISCES
27 APR–6 JUN	ARIES
6 JUN–17 JUL	TAURUS
17 JUL–1 SEP	GEMINI
1 SEP–26 OCT	CANCER
26 OCT–31 DEC	LEO

1977	Venus in
1 JAN–4 JAN	AQUARIUS
4 JAN–2 FEB	PISCES
2 FEB–6 JUN	ARIES
6 JUN–6 JUL	TAURUS
6 JUL–2 AUG	GEMINI
2 AUG–28 AUG	CANCER
28 AUG–22 SEP	LEO
22 SEP–17 OCT	VIRGO
17 OCT–10 NOV	LIBRA
10 NOV–4 DEC	SCORPIO
4 DEC–27 DEC	SAGITTARIUS
27 DEC–31 DEC	CAPRICORN

[illegible]	Mars in
[illegible]	[illegible]
12 DEC–31 DEC	[illegible]

1978	Venus in
[illegible] JAN–20 JAN	CAPRICORN
[illegible]	[illegible]ARIUS
12 JU[illegible]	[illegible]
8 AUG–7 SEP	LIBRA
7 SEP–31 DEC	SCORPIO

1979	Mars in
1 JAN–20 JAN	CAPRICORN
20 JAN–27 FEB	AQUARIUS
27 FEB–7 APR	PISCES
7 APR–16 MAY	ARIES
16 MAY–26 JUN	TAURUS
26 JUN–8 AUG	GEMINI
8 AUG–24 SEP	CANCER
24 SEP–19 NOV	LEO
19 NOV–31 DEC	VIRGO

1979	Venus in
1 JAN–7 JAN	SCORPIO
7 JAN–5 FEB	SAGITTARIUS
5 FEB–3 MAR	CAPRICORN
3 MAR–29 MAR	AQUARIUS
29 MAR–23 APR	PISCES
23 APR–18 MAY	ARIES
18 MAY–11 JUN	TAURUS
11 JUN–6 JUL	GEMINI
6 JUL–30 JUL	CANCER
30 JUL–24 AUG	LEO
24 AUG–17 SEP	VIRGO
17 SEP–11 OCT	LIBRA
11 OCT–4 NOV	SCORPIO
4 NOV–28 NOV	SAGITTARIUS
28 NOV–22 DEC	CAPRICORN
22 DEC–31 DEC	AQUARIUS

1980	Mars in
1 JAN–11 MAR	VIRGO
11 MAR–4 MAY	LEO
4 MAY–10 JUL	VIRGO
10 JUL–29 AUG	LIBRA
29 AUG–12 OCT	SCORPIO
12 OCT–22 NOV	SAGITTARIUS
22 NOV–30 DEC	CAPRICORN
30 DEC–31 DEC	AQUARIUS

1980	Venus in
1 JAN–16 JAN	AQUARIUS
16 JAN–9 FEB	PISCES
9 FEB–6 MAR	ARIES
6 MAR–3 APR	TAURUS
3 APR–12 MAY	GEMINI
12 MAY–5 JUN	CANCER
5 JUN–6 AUG	GEMINI
6 AUG–7 SEP	CANCER
7 SEP–4 OCT	LEO
4 OCT–30 OCT	VIRGO
30 OCT–24 NOV	LIBRA
24 NOV–18 DEC	SCORPIO
18 DEC–31 DEC	SAGITTARIUS

1981	Mars in
1 JAN–6 FEB	AQUARIUS
6 FEB–17 MAR	PISCES
17 MAR–25 APR	ARIES
25 APR–5 JUN	TAURUS
5 JUN–18 JUL	GEMINI
18 JUL–2 SEP	CANCER
2 SEP–21 OCT	LEO
21 OCT–16 DEC	VIRGO
16 DEC–31 DEC	LIBRA

1981	Venus in
1 JAN–11 JAN	SAGITTARIUS
11 JAN–4 FEB	CAPRICORN
4 FEB–28 FEB	AQUARIUS
28 FEB–24 MAR	PISCES
24 MAR–17 APR	ARIES
17 APR–11 MAY	TAURUS
11 MAY–5 JUN	GEMINI
5 JUN–29 JUN	CANCER
29 JUN–24 JUL	LEO
24 JUL–18 AUG	VIRGO
18 AUG–12 SEP	LIBRA
12 SEP–9 OCT	SCORPIO
9 OCT–5 NOV	SAGITTARIUS
5 NOV–8 DEC	CAPRICORN
8 DEC–31 DEC	AQUARIUS

1982	Venus in
[illegible]	AQUARIUS
14 AUG–7 SEP	[illegible]
7 SEP–2 OCT	VIRGO
2 OCT–26 OCT	LIBRA
26 OCT–18 NOV	SCORPIO
18 NOV–12 DEC	SAGITTARIUS
12 DEC–31 DEC	CAPRICORN

1983	Mars in
1 JAN–17 JAN	AQUARIUS
17 JAN–25 FEB	PISCES
25 FEB–5 APR	ARIES
5 APR–16 MAY	TAURUS
16 MAY–29 JUN	GEMINI
29 JUN–13 AUG	CANCER
13 AUG–30 SEP	LEO
30 SEP–18 NOV	VIRGO
18 NOV–31 DEC	LIBRA

1983	Venus in
1 JAN–5 JAN	CAPRICORN
5 JAN–29 JAN	AQUARIUS
29 JAN–22 FEB	PISCES
22 FEB–19 MAR	ARIES
19 MAR–13 APR	TAURUS
13 APR–9 MAY	GEMINI
9 MAY–6 JUN	CANCER
6 JUN–10 JUL	LEO
10 JUL–27 AUG	VIRGO
27 AUG–5 OCT	LEO
5 OCT–9 NOV	VIRGO
9 NOV–6 DEC	LIBRA
6 DEC–31 DEC	SCORPIO

1984	Mars in
1 JAN–11 JAN	LIBRA
11 JAN–17 AUG	SCORPIO
17 AUG–5 OCT	SAGITTARIUS
5 OCT–15 NOV	CAPRICORN
15 NOV–25 DEC	AQUARIUS
25 DEC–31 DEC	PISCES

1984	Venus in
1 JAN	SCORPIO
1 JAN–25 JAN	SAGITTARIUS
25 JAN–19 FEB	CAPRICORN
19 FEB–14 MAR	AQUARIUS
14 MAR–7 APR	PISCES
7 APR–2 MAY	ARIES
2 MAY–26 MAY	TAURUS
26 MAY–20 JUN	GEMINI
20 JUN–14 JUL	CANCER
14 JUL–7 AUG	LEO
7 AUG–1 SEP	VIRGO
1 SEP–25 SEP	LIBRA
25 SEP–20 OCT	SCORPIO
20 OCT–13 NOV	SAGITTARIUS
13 NOV–9 DEC	CAPRICORN
9 DEC–31 DEC	AQUARIUS

1985	Mars in
1 JAN–2 FEB	PISCES
2 FEB–15 MAR	ARIES
15 MAR–26 APR	TAURUS
26 APR–9 JUN	GEMINI
9 JUN–25 JUL	CANCER
25 JUL–10 SEP	LEO
10 SEP–27 OCT	VIRGO
27 OCT–14 DEC	LIBRA
14 DEC–31 DEC	SCORPIO

1985	Venus in
1 JAN–4 JAN	AQUARIUS
4 JAN–2 FEB	PISCES
2 FEB–6 JUN	ARIES
6 JUN–6 JUL	TAURUS
6 JUL–2 AUG	GEMINI
2 AUG–28 AUG	CANCER
28 AUG–22 SEP	LEO
22 SEP–16 OCT	VIRGO
16 OCT–9 NOV	LIBRA
9 NOV–3 DEC	SCORPIO
3 DEC–27 DEC	SAGITTARIUS
27 DEC–31 DEC	CAPRICORN

1986	Venus in
[illegible]	CAPRICORN
[illegible]	[illegible]
11 JUL–7 [illegible]	[illegible]
7 AUG–7 SEP	LIBRA
7 SEP–31 DEC	SCORPIO

1987	Mars in
1 JAN–8 JAN	PISCES
8 JAN–20 FEB	ARIES
20 FEB–5 APR	TAURUS
5 APR–21 MAY	GEMINI
21 MAY–6 JUL	CANCER
6 JUL–22 AUG	LEO
22 AUG–8 OCT	VIRGO
8 OCT–24 NOV	LIBRA
24 NOV–31 DEC	SCORPIO

1987	Venus in
1 JAN–7 JAN	SCORPIO
7 JAN–5 FEB	SAGITTARIUS
5 FEB–3 MAR	CAPRICORN
3 MAR–28 MAR	AQUARIUS
28 MAR–22 APR	PISCES
22 APR–17 MAY	ARIES
17 MAY–11 JUN	TAURUS
11 JUN–5 JUL	GEMINI
5 JUL–30 JUL	CANCER
30 JUL–23 AUG	LEO
23 AUG–16 SEP	VIRGO
16 SEP–10 OCT	LIBRA
10 OCT–3 NOV	SCORPIO
3 NOV–28 NOV	SAGITTARIUS
28 NOV–22 DEC	CAPRICORN
22 DEC–31 DEC	AQUARIUS

1988	Mars in
1 JAN–8 JAN	SCORPIO
8 JAN–22 FEB	SAGITTARIUS
22 FEB–6 APR	CAPRICORN
6 APR–22 MAY	AQUARIUS
22 MAY–13 JUL	PISCES
13 JUL–23 OCT	ARIES
23 OCT–1 NOV	PISCES
1 NOV–31 DEC	ARIES

1988	Venus in
1 JAN–15 JAN	AQUARIUS
15 JAN–9 FEB	PISCES
9 FEB–6 MAR	ARIES
6 MAR–3 APR	TAURUS
3 APR–17 MAY	GEMINI
17 MAY–27 MAY	CANCER
27 MAY–6 AUG	GEMINI
6 AUG–7 SEP	CANCER
7 SEP–4 OCT	LEO
4 OCT–29 OCT	VIRGO
29 OCT–23 NOV	LIBRA
23 NOV–17 DEC	SCORPIO
17 DEC–31 DEC	SAGITTARIUS

1989	Mars in
1 JAN–19 JAN	ARIES
19 JAN–11 MAR	TAURUS
11 MAR–29 APR	GEMINI
29 APR–16 JUN	CANCER
16 JUN–3 AUG	LEO
3 AUG–19 SEP	VIRGO
19 SEP–4 NOV	LIBRA
4 NOV–18 DEC	SCORPIO
18 DEC–31 DEC	SAGITTARIUS

1989	Venus in
1 JAN–10 JAN	SAGITTARIUS
10 JAN–3 FEB	CAPRICORN
3 FEB–27 FEB	AQUARIUS
27 FEB–23 MAR	PISCES
23 MAR–16 APR	ARIES
16 APR–11 MAY	TAURUS
11 MAY–4 JUN	GEMINI
4 JUN–29 JUN	CANCER
29 JUN–24 JUL	LEO
24 JUL–18 AUG	VIRGO
18 AUG–12 SEP	LIBRA
12 SEP–8 OCT	SCORPIO
8 OCT–5 NOV	SAGITTARIUS
5 NOV–10 DEC	CAPRICORN
10 DEC–31 DEC	AQUARIUS

[illegible]	Mars in
[illegible]	[illegible]
14 DEC–31 DEC	TAU[illegible]

1990	Venus in
[illegible] JAN	AQUARIUS
[illegible]	[illegible]
13 AUG–[illegible]	[illegible]
7 SEP–1 OCT	VIRGO
1 OCT–25 OCT	LIBRA
25 OCT–18 NOV	SCORPIO
18 NOV–12 DEC	SAGITTARIUS
12 DEC–31 DEC	CAPRICORN

1991	Mars in
1 JAN–21 JAN	TAURUS
21 JAN–3 APR	GEMINI
3 APR–26 MAY	CANCER
26 MAY–15 JUL	LEO
15 JUL–1 SEP	VIRGO
1 SEP–16 OCT	LIBRA
16 OCT–29 NOV	SCORPIO
29 NOV–31 DEC	SAGITTARIUS

1991	Venus in
1 JAN–5 JAN	CAPRICORN
5 JAN–29 JAN	AQUARIUS
29 JAN–22 FEB	PISCES
22 FEB–18 MAR	ARIES
18 MAR–13 APR	TAURUS
13 APR–9 MAY	GEMINI
9 MAY – 6 JUN	CANCER
6 JUN–11 JUL	LEO
11 JUL–21 AUG	VIRGO
21 AUG–6 OCT	LEO
6 OCT–9 NOV	VIRGO
9 NOV–6 DEC	LIBRA
6 DEC–31 DEC	SCORPIO
31 DEC	SAGITTARIUS

1992	Mars in
1 JAN–9 JAN	SAGITTARIUS
9 JAN–18 FEB	CAPRICORN
18 FEB–28 MAR	AQUARIUS
28 MAR–5 MAY	PISCES
5 MAY–14 JUN	ARIES
14 JUN–26 JUL	TAURUS
26 JUL–12 SEP	GEMINI
12 SEP–31 DEC	CANCER

1992	Venus in
1 JAN–25 JAN	SAGITTARIUS
25 JAN–18 FEB	CAPRICORN
18 FEB–13 MAR	AQUARIUS
13 MAR–7 APR	PISCES
7 APR–1 MAY	ARIES
1 MAY–26 MAY	TAURUS
26 MAY–19 JUN	GEMINI
19 JUN–13 JUL	CANCER
13 JUL–7 AUG	LEO
7 AUG–31 AUG	VIRGO
31 AUG–25 SEP	LIBRA
25 SEP–19 OCT	SCORPIO
19 OCT–13 NOV	SAGITTARIUS
13 NOV–8 DEC	CAPRICORN
8 DEC–31 DEC	AQUARIUS

1993	Mars in
1 JAN–27 APR	CANCER
27 APR–23 JUN	LEO
23 JUN–12 AUG	VIRGO
12 AUG–27 SEP	LIBRA
27 SEP–9 NOV	SCORPIO
9 NOV–20 DEC	SAGITTARIUS
20 DEC–31 DEC	CAPRICORN

1993	Venus in
1 JAN–3 JAN	AQUARIUS
3 JAN–2 FEB	PISCES
2 FEB–6 JUN	ARIES
6 JUN–6 JUL	TAURUS
6 JUL–1 AUG	GEMINI
1 AUG–27 AUG	CANCER
27 AUG–21 SEP	LEO
21 SEP–16 OCT	VIRGO
16 OCT–9 NOV	LIBRA
9 NOV–2 DEC	SCORPIO
2 DEC–26 DEC	SAGITTARIUS
26 DEC–31 DEC	CAPRICORN

1994	Mars in
[illegible]	CAPRICORN
12 DEC–31 DEC	[illegible]

1994	Venus in
1 JAN–19 JAN	CAPRICORN
[illegible]	[illegible]
7 AUG–7 SEP	LIBRA
7 SEP–31 DEC	SCORPIO

1995	Mars in
1 JAN–22 JAN	VIRGO
22 JAN–25 MAY	LEO
25 MAY–21 JUL	VIRGO
21 JUL–7 SEP	LIBRA
7 SEP–20 OCT	SCORPIO
20 OCT–30 NOV	SAGITTARIUS
30 NOV–31 DEC	CAPRICORN

1995	Venus in
1 JAN–7 JAN	SCORPIO
7 JAN–4 FEB	SAGITTARIUS
4 FEB–2 MAR	CAPRICORN
2 MAR–28 MAR	AQUARIUS
28 MAR–22 APR	PISCES
22 APR–16 MAY	ARIES
16 MAY–10 JUN	TAURUS
10 JUN–5 JUL	GEMINI
5 JUL–29 JUL	CANCER
29 JUL–23 AUG	LEO
23 AUG–16 SEP	VIRGO
16 SEP–10 OCT	LIBRA
10 OCT–3 NOV	SCORPIO
3 NOV–27 NOV	SAGITTARIUS
27 NOV–21 DEC	CAPRICORN
21 DEC–31 DEC	AQUARIUS

1996	Mars in
1 JAN–8 JAN	CAPRICORN
8 JAN–15 FEB	AQUARIUS
15 FEB–24 MAR	PISCES
24 MAR–2 MAY	ARIES
2 MAY–12 JUN	TAURUS
12 JUN–25 JUL	GEMINI
25 JUL–9 SEP	CANCER
9 SEP–30 OCT	LEO
30 OCT–31 DEC	VIRGO

1996	Venus in
1 JAN–15 JAN	AQUARIUS
15 JAN–9 FEB	PISCES
9 FEB–6 MAR	ARIES
6 MAR–3 APR	TAURUS
3 APR–7 AUG	GEMINI
7 AUG–7 SEP	CANCER
7 SEP–4 OCT	LEO
4 OCT–29 OCT	VIRGO
29 OCT–23 NOV	LIBRA
23 NOV–17 DEC	SCORPIO
17 DEC–31 DEC	SAGITTARIUS

1997	Mars in
1 JAN–3 JAN	VIRGO
3 JAN–8 MAR	LIBRA
8 MAR–19 JUN	VIRGO
19 JUN–14 AUG	LIBRA
14 AUG–28 SEP	SCORPIO
28 SEP–9 NOV	SAGITTARIUS
9 NOV–18 DEC	CAPRICORN
18 DEC–31 DEC	AQUARIUS

1997	Venus in
1 JAN–10 JAN	SAGITTARIUS
10 JAN–3 FEB	CAPRICORN
3 FEB–27 FEB	AQUARIUS
27 FEB–23 MAR	PISCES
23 MAR–16 APR	ARIES
16 APR–10 MAY	TAURUS
10 MAY–4 JUN	GEMINI
4 JUN–28 JUN	CANCER
28 JUN–23 JUL	LEO
23 JUL–17 AUG	VIRGO
17 AUG–12 SEP	LIBRA
12 SEP–8 OCT	SCORPIO
8 OCT–5 NOV	SAGITTARIUS
5 NOV–12 DEC	CAPRICORN
12 DEC–31 DEC	AQUARIUS

1998	Mars in	1998	Venus in
1 JAN–25 JAN	AQUARIUS	1 JAN–9 JAN	AQUARIUS
		9 JAN–4 MAR	CAPRICORN
		30 SEP–24 OCT	LIBRA
		24 OCT–17 NOV	SCORPIO
		17 NOV–11 DEC	SAGITTARIUS
		11 DEC–31 DEC	CAPRICORN

1999	Mars in	1999	Venus in
1 JAN–26 JAN	LIBRA	1 JAN–4 JAN	CAPRICORN
26 JAN–5 MAY	SCORPIO	4 JAN–28 JAN	AQUARIUS
5 MAY–5 JUL	LIBRA	28 JAN–21 FEB	PISCES
5 JUL–2 SEP	SCORPIO	21 FEB–18 MAR	ARIES
2 SEP–17 OCT	SAGITTARIUS	18 MAR–12 APR	TAURUS
17 OCT–26 NOV	CAPRICORN	12 APR–8 MAY	GEMINI
26 NOV–31 DEC	AQUARIUS	8 MAY–5 JUN	CANCER
		5 JUN–12 JUL	LEO
		12 JUL–15 AUG	VIRGO
		15 AUG–7 OCT	LEO
		7 OCT–9 NOV	VIRGO
		9 NOV–5 DEC	LIBRA
		5 DEC–31 DEC	SCORPIO
		31 DEC	SAGITTARIUS

2000	Mars in
1 JAN–4 JAN	AQUARIUS
4 JAN–12 FEB	PISCES
12 FEB–23 MAR	ARIES
23 MAR–3 MAY	TAURUS
3 MAY–16 JUN	GEMINI
16 JUN–1 AUG	CANCER
1 AUG–17 SEP	LEO
17 SEP–4 NOV	VIRGO
4 NOV–23 DEC	LIBRA
23 DEC–31 DEC	SCORPIO

2000	Venus in
1 JAN–24 JAN	SAGITTARIUS
24 JAN–18 FEB	CAPRICORN
18 FEB–13 MAR	AQUARIUS
13 MAR–6 APR	PISCES
6 APR–1 MAY	ARIES
1 MAY–25 MAY	TAURUS
25 MAY–18 JUN	GEMINI
18 JUN–13 JUL	CANCER
13 JUL–6 AUG	LEO
6 AUG–31 AUG	VIRGO
31 AUG–24 SEP	LIBRA
24 SEP–19 OCT	SCORPIO
19 OCT–13 NOV	SAGITTARIUS
13 NOV–8 DEC	CAPRICORN
8 DEC–31 DEC	AQUARIUS

References

Quotations in this book were drawn from the following texts:

Bacall, L., *By Myself* (Coronet, London, 1979): pp. 26, 218.

Duncan, I., *My Life* (Sphere, Bucks., 1968): p. 179.

Fergusson, R., *The Penguin Dictionary of Proverbs* (Penguin, Middlesex, 1983): pp. 22, 187, 240.

Flynn, E., *My Wicked, Wicked Ways* (Pan, London, 1961): pp. 318, 323, 325.

Lesley, C., *The Life of Noel Coward* (Penguin, Middlesex, 1978): p. 89.

Levinson, L., The Left Handed Dictionary (Collier, New York, 1963): p. 85.

McClelland, D., *StarSpeak – Hollywood on Everything* (Faber & Faber, Boston, 1987): pp. 174, 273.

Neal, P., *As I am* (Arrow, London, 1989): p. 119.

Oxford Dictionary of Quotations (OUP, London, 1974): pp. 569, 573.

Pearson, H., *Bernard Shaw* (Reprint Society, London, 1948): pp. 160, 440.

Redgrave, V., *An Autobiography* (Arrow, London, 1992): p. 136.

Rose, P. (ed), *The Penguin Book of Women's Lives* (Viking, London, 1994): pp. 53, 619.

Shipman, D., *Movie Talk – Who Said What About Whom In The Movies* (Bloomsbury, London, 1988): pp. 8, 11, 56, 58, 88, 96, 100, 106, 116, 117, 121, 148, 153, 195, 212, 219.

Steiger, B. & Mank C. – *Valentino* (Corgi, London, 1976): p. 123.

The Age, Melbourne: 27.2.94, 25.6.95, 16.7.95, 7.4.96, 5.5.96, 1.12.96., 20.7.97.

Vanity Fair, January 1995.
Wallace, I., A. & S. & Wallechinsky, D., *The Intimate Sex Lives of Famous People* (Arrow, London, 1982): pp. 71, 79, 104, 180–183, 253, 360, 452–3.
Who Weekly: 3.7.95.
Wintle, J. & Kenin, R. (eds), *The Penguin Concise Dictionary of Biographical Quotation* (Penguin, Middlesex, 1981): pp. 220, 322.